# PARTICLE PHYSICS

Edited by **Eugene Kennedy**

**Particle Physics**
Edited by Eugene Kennedy

**Published by InTech**
Janeza Trdine 9, 51000 Rijeka, Croatia



**Publishing Process Manager** Vana Persen

**Technical Editor** Teodora Smiljanic

**Cover Designer** InTech Design Team

Additional hard copies can be obtained from orders@intechopen.com

Particle Physics, Edited by Eugene Kennedy
p. cm.
ISBN 978-953-51-0481-0

For EU product safety concerns:
IN TECH d.o.o., Prolaz Marije Krucifikse Kozulić 3, 51000 Rijeka, Croatia,
info@intechopen.com or visit our website at intechopen.com.

# Meet the editor

Professor Kennedy was appointed in 1986 as Professor of Physics and Head of School at Dublin City University (DCU) and served as Dean of the Faculty of Science from 1987 to 1994. He has published widely, served on international Advisory Boards, coordinated EU research networks and carried out research at facilities in Wash. DC, Hamburg, Paris and Berkeley. He was the founding Director of the National Centre for Plasma Science and Technology and was appointed university Vice-President for Research in December 2004, holding this position until his recent retirement. He served as Chair of the Radiological Protection Institute of Ireland from 2006 to 2011. He is now Professor Emeritus at DCU.

# Contents

# Preface

Interest in particle physics continues apace. With the Large Hadron Collider showing early tantalizing glimpses of what may yet prove to be the elusive Higgs Boson, particle physics remains a fertile ground for creative theorists. While the Standard model of particle physics remains hugely successful, nevertheless it is still not fully regarded as a complete holistic description. This book describes the development of what is termed the generation model, which is proposed as an alternative to the standard model and provides a new classification approach to fundamental particles. A further chapter describes an extension to the standard model involving the possibility of a charged Higgs boson and includes an outline of how experimental evidence may be sought at LHC and B-factory facilities. Coupling of postulated axion particles to photons is tackled with particular reference to magnetized media, together with possible implications for detection in laboratory experiments or astrophysical observations. Modern particle physics now involves major investments in hardware coupled with large-scale theoretical and computational efforts. The complexity of such synergistic coordinated entities is illustrated within the framework of the e-science paradigm. Finally, an unexpected and interesting description of the potential radiation hazards associated with extremely weakly interacting neutrinos is provided in the context of possible future designs of intense muon-collider facilities.

**Eugene Kennedy**
Emeritus Professor
School of Physical Sciences,
Dublin City University
Ireland

1

# The Generation Model of Particle Physics

Brian Robson
*Department of Theoretical Physics, Research School of Physics and Engineering, The Australian National University, Canberra Australia*

## 1. Introduction

The main purpose of this chapter is to present an alternative to the Standard Model (SM) (Gottfried and Weisskopf, 1984) of particle physics. This alternative model, called the Generation Model (GM) (Robson, 2002; 2004; Evans and Robson, 2006), describes all the transition probabilities for interactions involving the six leptons and the six quarks, which form the elementary particles of the SM in terms of only three unified additive quantum numbers instead of the nine non-unified additive quantum numbers allotted to the leptons and quarks in the SM.

The chapter presents (Section 2) an outline of the current formulation of the SM: the elementary particles and the fundamental interactions of the SM, and the basic problem inherent in the SM. This is followed by (Section 3) a summary of the GM, highlighting the essential differences between the GM and the SM. Section 3 also introduces a more recent development of a composite GM in which both leptons and quarks have a substructure. This enhanced GM has been named the Composite Generation Model (CGM) (Robson, 2005; 2011a). In this chapter, for convenience, we shall refer to this enhanced GM as the CGM, whenever the substructure of leptons and quarks is important for the discussion. Section 4 focuses on several important consequences of the different paradigms provided by the GM. In particular: the origin of mass, the mass hierarchy of the leptons and quarks, the origin of gravity and the origin of apparent CP violation, are discussed. Finally, Section 5 provides a summary and discusses future prospects.

## 2. Standard model of particle physics

The Standard Model (SM) of particle physics (Gottfried and Weisskopf, 1984) was developed throughout the 20th century, although the current formulation was essentially finalized in the mid-1970s following the experimental confirmation of the existence of quarks (Bloom *et al.*, 1969; Breidenbach *et al.*, 1969).

The SM has enjoyed considerable success in describing the interactions of leptons and the multitude of hadrons (baryons and mesons) with each other as well as the decay modes of the unstable leptons and hadrons. However the model is considered to be incomplete in the sense that it provides no understanding of several empirical observations such as: the existence of three families or generations of leptons and quarks, which apart from mass have similar properties; the mass hierarchy of the elementary particles, which form the basis of the SM; the nature of the gravitational interaction and the origin of CP violation.

In this section a summary of the current formulation of the SM is presented: the elementary particles and the fundamental interactions of the SM, and then the basic problem inherent in the SM.

### 2.1 Elementary particles of the SM

In the SM the elementary particles that are the constituents of matter are assumed to be the six leptons: electron neutrino ($\nu_e$), electron ($e^-$), muon neutrino ($\nu_\mu$), muon ($\mu^-$), tau neutrino ($\nu_\tau$), tau ($\tau^-$) and the six quarks: up ($u$), down ($d$), charmed ($c$), strange ($s$), top ($t$) and bottom ($b$), together with their antiparticles. These twelve particles are all spin-$\frac{1}{2}$ particles and fall naturally into three families or generations: (i) $\nu_e, e^-, u, d$ ; (ii) $\nu_\mu, \mu^-, c, s$ ; (iii) $\nu_\tau, \tau^-, t, b$ . Each generation consists of two leptons with charges $Q = 0$ and $Q = -1$ and two quarks with charges $Q = +\frac{2}{3}$ and $Q = -\frac{1}{3}$. The masses of the particles increase significantly with each generation with the possible exception of the neutrinos, whose very small masses have yet to be determined.

In the SM the leptons and quarks are allotted several additive quantum numbers: charge $Q$, lepton number $L$, muon lepton number $L_\mu$, tau lepton number $L_\tau$, baryon number $A$, strangeness $S$, charm $C$, bottomness $B$ and topness $T$. These are given in Table 1. For each particle additive quantum number $N$, the corresponding antiparticle has the additive quantum number $-N$.

| particle | $Q$ | $L$ | $L_\mu$ | $L_\tau$ | $A$ | $S$ | $C$ | $B$ | $T$ |
|---|---|---|---|---|---|---|---|---|---|
| $\nu_e$ | 0 | 1 | 0 | 0 | 0 | 0 | 0 | 0 | 0 |
| $e^-$ | $-1$ | 1 | 0 | 0 | 0 | 0 | 0 | 0 | 0 |
| $\nu_\mu$ | 0 | 1 | 1 | 0 | 0 | 0 | 0 | 0 | 0 |
| $\mu^-$ | $-1$ | 1 | 1 | 0 | 0 | 0 | 0 | 0 | 0 |
| $\nu_\tau$ | 0 | 1 | 0 | 1 | 0 | 0 | 0 | 0 | 0 |
| $\tau^-$ | $-1$ | 1 | 0 | 1 | 0 | 0 | 0 | 0 | 0 |
| $u$ | $+\frac{2}{3}$ | 0 | 0 | 0 | $\frac{1}{3}$ | 0 | 0 | 0 | 0 |
| $d$ | $-\frac{1}{3}$ | 0 | 0 | 0 | $\frac{1}{3}$ | 0 | 0 | 0 | 0 |
| $c$ | $+\frac{2}{3}$ | 0 | 0 | 0 | $\frac{1}{3}$ | 0 | 1 | 0 | 0 |
| $s$ | $-\frac{1}{3}$ | 0 | 0 | 0 | $\frac{1}{3}$ | $-1$ | 0 | 0 | 0 |
| $t$ | $+\frac{2}{3}$ | 0 | 0 | 0 | $\frac{1}{3}$ | 0 | 0 | 0 | 1 |
| $b$ | $-\frac{1}{3}$ | 0 | 0 | 0 | $\frac{1}{3}$ | 0 | 0 | $-1$ | 0 |

Table 1. SM additive quantum numbers for leptons and quarks

Table 1 demonstrates that, except for charge, leptons and quarks are allotted different kinds of additive quantum numbers so that this classification of the elementary particles in the SM is *non-unified*.

The additive quantum numbers $Q$ and $A$ are assumed to be conserved in strong, electromagnetic and weak interactions. The lepton numbers $L$, $L_\mu$ and $L_\tau$ are not involved in strong interactions but are strictly conserved in both electromagnetic and weak interactions. The remainder, $S$, $C$, $B$ and $T$ are strictly conserved only in strong and electromagnetic interactions but can undergo a change of one unit in weak interactions.

The quarks have an additional additive quantum number called "color charge", which can take three values so that in effect we have three kinds of each quark, $u$, $d$, etc. These are often called red, green and blue quarks. The antiquarks carry anticolors, which for simplicity are called antired, antigreen and antiblue. Each quark or antiquark carries a single unit of color or anticolor charge, respectively. The leptons do not carry a color charge and consequently do not participate in the strong interactions, which occur between particles carrying color charges.

### 2.2 Fundamental interactions of the SM

The SM recognizes four fundamental interactions in nature: strong, electromagnetic, weak and gravity. Since gravity plays no role in particle physics because it is so much weaker than the other three fundamental interactions, the SM does not attempt to explain gravity. In the SM the other three fundamental interactions are assumed to be associated with a local gauge field.

#### 2.2.1 Strong interactions

The strong interactions, mediated by massless neutral spin-1 gluons between quarks carrying a color charge, are described by an $SU(3)$ local gauge theory called quantum chromodynamics (QCD) (Halzen and Martin, 1984). There are eight independent kinds of gluons, each of which carries a combination of a color charge and an anticolor charge (e.g. red-antigreen). The strong interactions between color charges are such that in nature the quarks (antiquarks) are grouped into composites of either three quarks (antiquarks), called baryons (antibaryons), each having a different color (anticolor) charge or a quark-antiquark pair, called mesons, of opposite color charges. In the $SU(3)$ color gauge theory each baryon, antibaryon or meson is colorless. However, these colorless particles, called hadrons, may interact strongly via residual strong interactions arising from their composition of colored quarks and/or antiquarks. On the other hand the colorless leptons are assumed to be structureless in the SM and consequently do not participate in strong interactions.

#### 2.2.2 Electromagnetic interactions

The electromagnetic interactions, mediated by massless neutral spin-1 photons between electrically charged particles, are described by a $U(1)$ local gauge theory called quantum electrodynamics (Halzen and Martin, 1984).

#### 2.2.3 Weak interactions

The weak interactions, mediated by the massive $W^+$, $W^-$ and $Z^0$ vector bosons between all the elementary particles of the SM, fall into two classes: (i) charge-changing (CC) weak interactions involving the $W^+$ and $W^-$ bosons and (ii) neutral weak interactions involving the $Z^0$ boson. The CC weak interactions, acting exclusively on left-handed particles and right-handed antiparticles, are described by an $SU(2)_L$ local gauge theory, where the subscript $L$ refers to left-handed particles only (Halzen and Martin, 1984). On the other hand, the neutral weak interactions act on both left-handed and right-handed particles, similar to the electromagnetic interactions. In fact the SM assumes (Glashow, 1961) that both the $Z^0$ and the photon ($\gamma$) arise from a mixing of two bosons, $W^0$ and $B^0$, via an electroweak mixing angle $\theta_W$:

$$\gamma = B^0 \cos\theta_W + W^0 \sin\theta_W \,, \tag{1}$$

$$Z^0 = -B^0 \sin\theta_W + W^0 \cos\theta_W \,. \tag{2}$$

These are described by a $U(1) \times SU(2)_L$ local gauge theory, where the $U(1)$ symmetry involves both left-handed and right-handed particles.

Experiment requires the masses of the weak gauge bosons, $W$ and $Z$, to be heavy so that the weak interactions are very short-ranged. On the other hand, Glashow's proposal, based upon the concept of a non-Abelian $SU(2)$ Yang-Mills gauge theory, requires the mediators of the weak interactions to be massless like the photon. This boson mass problem was resolved by Weinberg (1967) and Salam (1968), who independently employed the idea of spontaneous symmetry breaking involving the Higgs mechanism (Englert and Brout, 1964; Higgs, 1964). In this way the $W$ and $Z$ bosons acquire mass and the photon remains massless.

The above treatment of the electromagnetic and weak interactions in terms of a $U(1) \times SU(2)_L$ local gauge theory has become known as the Glashow, Weinberg and Salam (GWS) model and forms one of the cornerstones of the SM. The model gives the relative masses of the $W$ and $Z$ bosons in terms of the electroweak mixing angle:

$$M_W = M_Z \cos\theta_W \,. \tag{3}$$

The Higgs mechanism was also able to cure the associated fermion mass problem (Aitchison and Hey, 1982): the finite masses of the leptons and quarks cause the Lagrangian describing the system to violate the $SU(2)_L$ gauge invariance. By coupling originally massless fermions to a scalar Higgs field, it is possible to produce the observed physical fermion masses without violating the gauge invariance. However, the GWS model requires the existence of a new massive spin zero boson, the Higgs boson, which to date remains to be detected. In addition, the fermion-Higgs coupling strength is dependent upon the mass of the fermion so that a new parameter is required for each fermion mass in the theory.

In 1971, t'Hooft (1971a,b) showed that the GWS model of the electroweak interactions was renormalizable and this self-consistency of the theory led to its general acceptance. In 1973, events corresponding to the predicted neutral currents mediated by the $Z^0$ boson were observed (Hasert *et al.*, 1973; 1974), while bosons, with approximately the expected masses, were discovered in 1983 (Arnison *et al.*, 1983; Banner *et al.*, 1983), thereby confirming the GWS model.

Another important property of the CC weak interactions is their universality for both leptonic and hadronic processes. In the SM this property is taken into account differently for leptonic and hadronic processes.

For leptonic CC weak interaction processes, each of the charged leptons is assumed to form a weak isospin doublet ($i = \frac{1}{2}$) with its respective neutrino, i.e. $(\nu_e, e^-)$, $(\nu_\mu, \mu^-)$, $(\nu_\tau, \tau^-)$, with each doublet having the third component of weak isospin $i_3 = (+\frac{1}{2}, -\frac{1}{2})$. In addition each doublet is associated with a different lepton number so that there are no CC weak interaction transitions between generations. Thus for leptonic processes, the concept of a universal CC weak interaction allows one to write (for simplicity we restrict the discussion to the first two generations only):

$$a(\nu_e, e^-; W^-) = a(\nu_\mu, \mu^-; W^-) = g_w \,. \tag{4}$$

Here $a(\alpha, \beta; W^-)$ represents the CC weak interaction transition amplitude involving the fermions $\alpha, \beta$ and the $W^-$ boson, and $g_w$ is the universal CC weak interaction transition amplitude. Lepton number conservation gives

$$a(\nu_e, \mu^-; W^-) = a(\nu_\mu, e^-; W^-) = 0\,, \tag{5}$$

so that there are no CC weak interaction transitions between generations in agreement with experiment.

Unlike the pure leptonic decays, which are determined by the conservation of the various lepton numbers, there is no quantum number in the SM which restricts quark (hadronic) CC weak interaction processes between generations. In the SM the quarks do not appear to form weak isospin doublets: the known decay processes of neutron $\beta$-decay and $\Lambda^0$ $\beta$-decay suggest that quarks mix between generations and that the "flavor" quantum numbers, $S$, $C$, $B$ and $T$ are not necessarily conserved in CC weak interaction processes.

In the SM neutron $\beta$-decay:

$$n^0 \to p^+ + e^- + \bar{\nu}_e\,, \tag{6}$$

is interpreted as the sequential transition

$$d \to u + W^-\,, \qquad W^- \to e^- + \bar{\nu}_e\,. \tag{7}$$

The overall coupling strength of the CC weak interactions involved in neutron $\beta$-decay was found to be slightly weaker ($\approx 0.95$) than that for muon decay:

$$\mu^- \to \nu_\mu + W^-\,, \qquad W^- \to e^- + \bar{\nu}_e\,. \tag{8}$$

Similarly, $\Lambda^0$ $\beta$-decay:

$$\Lambda^0 \to p^+ + e^- + \bar{\nu}_e\,, \tag{9}$$

is interpreted in the SM as the sequential transition

$$s \to u + W^-\,, \qquad W^- \to e^- + \bar{\nu}_e\,. \tag{10}$$

In this case the overall coupling strength of the CC weak interactions was found to be significantly less ($\approx 0.05$) than that for muon decay.

In the SM the universality of the CC weak interaction for both leptonic and hadronic processes is restored by adopting the proposal of Cabibbo (1963) that in hadronic processes the CC weak interaction is *shared* between $\Delta S = 0$ and $\Delta S = 1$ transition amplitudes in the ratio of $\cos\theta_c$ : $\sin\theta_c$. The Cabibbo angle $\theta_c$ has a value $\approx 13^0$, which gives good agreement with experiment for the decay processes (7) and (10) relative to (8).

This "Cabibbo mixing" is an integral part of the SM. In the quark model it leads to a sharing of the CC weak interaction between quarks with different flavors (different generations) unlike the corresponding case of leptonic processes. Again, in order to simplify matters, the following discussion (and also throughout the chapter) will be restricted to the first two generations of the elementary particles of the SM, involving only the Cabibbo mixing, although the extension to three generations is straightforward (Kobayashi and Maskawa, 1973). In the latter case, the quark mixing parameters correspond to the so-called Cabibbo-Kobayashi-Maskawa (CKM) matrix elements, which indicate that inclusion of the third generation would have a minimal effect on the overall coupling strength of the CC weak interactions.

Cabibbo mixing was incorporated into the quark model of hadrons by postulating that the so-called weak interaction eigenstate quarks, $d'$ and $s'$, form CC weak interaction isospin doublets with the $u$ and $c$ quarks, respectively: $(u, d')$ and $(c, s')$. These weak eigenstate quarks are linear superpositions of the so-called mass eigenstate quarks ($d$ and $s$):

$$d' = d \cos\theta_c + s \sin\theta_c \tag{11}$$

and

$$s' = -d \sin\theta_c + s \cos\theta_c\,. \tag{12}$$

The quarks $d$ and $s$ are the quarks which participate in the electromagnetic and the strong interactions with the full allotted strengths of electric charge and color charge, respectively. The quarks $d'$ and $s'$ are the quarks which interact with the $u$ and $c$ quarks, respectively, with the full strength of the CC weak interaction.

In terms of transition amplitudes, Eqs. (11) and (12) can be represented as

$$a(u, d'; W^-) = a(u, d; W^-) \cos\theta_c + a(u, s; W^-) \sin\theta_c = g_w \tag{13}$$

and

$$a(c, s'; W^-) = -a(c, d; W^-) \sin\theta_c + a(c, s; W^-) \cos\theta_c = g_w\,. \tag{14}$$

In addition one has the relations

$$a(u, s'; W^-) = -a(u, d; W^-) \sin\theta_c + a(u, s; W^-) \cos\theta_c = 0 \tag{15}$$

and

$$a(c, d'; W^-) = a(c, d; W^-) \cos\theta_c + a(c, s; W^-) \sin\theta_c = 0\,. \tag{16}$$

Eqs. (13) and (14) indicate that it is the $d'$ and $s'$ quarks which interact with the $u$ and $c$ quarks, respectively, with the full strength $g_w$. These equations for quarks correspond to Eq. (4) for leptons. Similarly, Eqs. (15) and (16) for quarks correspond to Eq. (5) for leptons. However, there is a fundamental difference between Eqs. (15) and (16) for quarks and Eq. (5) for leptons. The former equations do not yield zero amplitudes because there exists some quantum number (analagous to muon lepton number) which is required to be conserved. This lack of a selection rule indicates that the notion of weak isospin symmetry for the doublets $(u, d')$ and $(c, s')$ is dubious.

Eqs. (13) and (15) give

$$a(u, d; W^-) = g_w \cos\theta_c\,, \qquad a(u, s; W^-) = g_w \sin\theta_c\,. \tag{17}$$

Thus in the two generation approximation of the SM, transitions involving $d \to u + W^-$ proceed with a strength proportional to $g_w^2 \cos^2\theta_c \approx 0.95 g_w^2$, while transitions involving $s \to u + W^-$ proceed with a strength proportional to $g_w^2 \sin^2\theta_c \approx 0.05 g_w^2$, as required by experiment.

### 2.3 Basic problem inherent in SM

The basic problem with the SM is the classification of its elementary particles employing a diverse complicated scheme of additive quantum numbers (Table 1), some of which are not conserved in weak interaction processes; and at the same time failing to provide any physical basis for this scheme.

A good analogy of the SM situation is the Ptolemaic model of the universe, based upon a stationary Earth at the center surrounded by a rotating system of crystal spheres refined by the addition of epicycles (small circular orbits) to describe the peculiar movements of the planets around the Earth. While the Ptolemaic model yielded an excellent description, it is a complicated diverse scheme for predicting the movements of the Sun, Moon, planets and the stars around a stationary Earth and unfortunately provides no understanding of these complicated movements.

Progress in understanding the universe was only made when the Ptolemaic model was replaced by the Copernican-Keplerian model, in which the Earth moved like the other planets around the Sun, and Newton discovered his universal law of gravitation to describe the approximately elliptical planetary orbits.

The next section describes a new model of particle physics, the Generation Model (GM), which addresses the problem within the SM, replacing it with a much simpler and unified classification scheme of leptons and quarks, and providing some understanding of phenomena, which the SM is unable to address.

## 3. Generation model of particle physics

The Generation Model (GM) of particle physics has been developed over the last decade. In the initial paper (Robson, 2002) a new classification of the elementary particles, the six leptons and the six quarks, of the SM was proposed. This classification was based upon the use of only three additive quantum numbers: charge ($Q$), particle number ($p$) and generation quantum number ($g$), rather than the nine additive quantum numbers (see Table 1) of the SM. Thus the new classification is both simpler and unified in that leptons and quarks are assigned the same kind of additive quantum numbers unlike those of the SM. It will be discussed in more detail in Subsection 3.1.

Another feature of the new classification scheme is that all three additive quantum numbers, $Q$, $p$ and $g$, are required to be conserved in all leptonic and hadronic processes. In particular the generation quantum number $g$ is strictly conserved in weak interactions unlike some of the quantum numbers, e.g. strangeness $S$, of the SM. This latter requirement led to a new treatment of quark mixing in hadronic processes (Robson, 2002; Evans and Robson, 2006), which will be discussed in Subsection 3.2.

The development of the GM classification scheme, which provides a unified description of leptons and quarks, indicated that leptons and quarks are intimately related and led to the development of composite versions of the GM, which we refer to as the Composite Generation Model (CGM) (Robson, 2005; 2011a). The CGM will be discussed in Subsection 3.3.

Subsection 3.4 discusses the fundamental interactions of the GM.

### 3.1 Unified classification of leptons and quarks

Table 2 displays a set of three additive quantum numbers: charge ($Q$), particle number ($p$) and generation quantum number ($g$) for the unified classification of the leptons and quarks corresponding to the current CGM (Robson, 2011a). As for Table 1 the corresponding antiparticles have the opposite sign for each particle additive quantum number.

| particle | $Q$ | $p$ | $g$ | particle | $Q$ | $p$ | $g$ |
|---|---|---|---|---|---|---|---|
| $\nu_e$ | 0 | $-1$ | 0 | $u$ | $+\frac{2}{3}$ | $\frac{1}{3}$ | 0 |
| $e^-$ | $-1$ | $-1$ | 0 | $d$ | $-\frac{1}{3}$ | $\frac{1}{3}$ | 0 |
| $\nu_\mu$ | 0 | $-1$ | $\pm1$ | $c$ | $+\frac{2}{3}$ | $\frac{1}{3}$ | $\pm1$ |
| $\mu^-$ | $-1$ | $-1$ | $\pm1$ | $s$ | $-\frac{1}{3}$ | $\frac{1}{3}$ | $\pm1$ |
| $\nu_\tau$ | 0 | $-1$ | $0, \pm2$ | $t$ | $+\frac{2}{3}$ | $\frac{1}{3}$ | $0, \pm2$ |
| $\tau^-$ | $-1$ | $-1$ | $0, \pm2$ | $b$ | $-\frac{1}{3}$ | $\frac{1}{3}$ | $0, \pm2$ |

Table 2. CGM additive quantum numbers for leptons and quarks

Each generation of leptons and quarks has the same set of values for the additive quantum numbers $Q$ and $p$. The generations are differentiated by the generation quantum number $g$, which in general can have multiple values. The latter possibilities arise from the composite nature of the leptons and quarks in the CGM.

The three conserved additive quantum numbers, $Q$, $p$ and $g$ are sufficient to describe all the observed transition amplitudes for both hadronic and leptonic processes, provided each "force" particle, mediating the various interactions, has $p = g = 0$.

Comparison of Tables 1 and 2 indicates that the two models, SM and CGM, have only one additive quantum number in common, namely electric charge $Q$, which serves the same role in both models and is conserved. The second additive quantum number of the CGM, particle number $p$, replaces both lepton number $L$ and baryon number $A$ of the SM. The third additive quantum number of the CGM, generation quantum number $g$, effectively replaces the remaining additive quantum numbers of the SM, $L_\mu$, $L_\tau$, $S$, $C$, $B$ and $T$.

Table 2 shows that the CGM provides both a simpler and *unified* classification scheme for leptons and quarks. Furthermore, the generation quantum number $g$ is conserved in the CGM unlike the additive quantum numbers, $S$, $C$, $B$ and $T$ of the SM. Conservation of $g$ requires a new treatment of quark mixing in hadronic processes, which will be discussed in the next subsection.

### 3.2 Quark mixing in hadronic CC weak interaction processes in the GM

The GM differs from the SM in two fundamental ways, which are essential to preserve the universality of the CC weak interaction for both leptonic and hadronic processes. In the SM this was accomplished, initially by Cabibbo (1963) for the first two generations by the introduction of "Cabibbo quark mixing", and later by Kobayashi and Maskawa (1973), who generalized quark mixing involving the CKM matrix elements to the three generations.

Firstly, the GM postulates that the mass eigenstate quarks of the same generation, e.g. $(u, d)$, form weak isospin doublets and couple with the full strength of the CC weak interaction, $g_w$, like the lepton doublets, e.g. $(\nu_e, e^-)$. Unlike the SM, the GM requires that there is no coupling between mass eigenstate quarks from different generations. This latter requirement corresponds to the conservation of the generation quantum number $g$ in the CC weak interaction processes.

Secondly, the GM postulates that hadrons are composed of weak eigenstate quarks such as $d'$ and $s'$ given by Eqs. (11) and (12) in the two generation approximation, rather than the corresponding mass eigenstate quarks, $d$ and $s$, as in the SM.

To maintain lepton-quark universality for CC weak interaction processes in the two generation approximation, the GM postulates that

$$a(u,d;W^-) = a(c,s;W^-) = g_w \tag{18}$$

and generation quantum number conservation gives

$$a(u,s;W^-) = a(c,d;W^-) = 0\,. \tag{19}$$

Eqs. (18) and (19) are the analogues of Eqs. (4) and (5) for leptons. Thus the quark pairs $(u,d)$ and $(c,s)$ in the GM form weak isospin doublets, similar to the lepton pairs $(\nu_e, e^-)$ and $(\nu_\mu, \mu^-)$, thereby establishing a close lepton-quark parallelism with respect to weak isospin symmetry.

To account for the reduced transition probabilities for neutron and $\Lambda^0$ $\beta$-decays, the GM postulates that the neutron and $\Lambda^0$ baryon are composed of weak eigenstate quarks, $u$, $d'$ and $s'$. Thus, neutron $\beta$-decay is to be interpreted as the sequential transition

$$d' \to u + W^-\,, \qquad W^- \to e^- + \bar{\nu}_e\,. \tag{20}$$

The primary transition has the amplitude $a(u,d';W^-)$ given by

$$a(u,d';W^-) = a(u,d;W^-)\cos\theta_c + a(u,s;W^-)\sin\theta_c = g_w\cos\theta_c\,, \tag{21}$$

where we have used Eqs. (18) and (19). This gives the same transition probability for neutron $\beta$-decay $(g_w^4\cos^2\theta_c)$ relative to muon decay $(g_w^4)$ as the SM. Similarly, $\Lambda^0$ $\beta$-decay is to be interpreted as the sequential transition

$$s' \to u + W^-\,, \qquad W^- \to e^- + \bar{\nu}_e\,. \tag{22}$$

In this case the primary transition has the amplitude $a(u,s';W^-)$ given by

$$a(u,s';W^-) = -a(u,d;W^-)\sin\theta_c + a(u,s;W^-)\cos\theta_c = -g_w\sin\theta_c\,. \tag{23}$$

Thus $\Lambda^0$ $\beta$-decay has the same transition probability $(g_w^4\sin^2\theta_c)$ relative to muon decay $(g_w^4)$ as that given by the SM.

The GM differs from the SM in that it treats quark mixing differently from the method introduced by Cabibbo (1963) and employed in the SM. Essentially, in the GM, the quark mixing is placed in the quark states (wave functions) rather than in the CC weak interactions. This allows a unified and simpler classification of both leptons and quarks in terms of only three additive quantum numbers, $Q$, $p$ and $g$, each of which is conserved in all interactions.

### 3.3 Composite generation model

The unified classification scheme of the GM makes feasible a composite version of the GM (CGM) (Robson, 2005). This is not possible in terms of the non-unified classification scheme of the SM, involving different additive quantum numbers for leptons than for quarks and the non-conservation of some additive quantum numbers, such as strangeness, in the case of quarks. Here we shall present the current version (Robson, 2011a), which takes into account the mass hierarchy of the three generations of leptons and quarks. There is evidence that leptons and quarks, which constitute the elementary particles of the SM, are actually composites.

Firstly, the electric charges of the electron and proton are opposite in sign but are *exactly* equal in magnitude so that atoms with the same number of electrons and protons are neutral. Consequently, in a proton consisting of quarks, the electric charges of the quarks are intimately related to that of the electron: in fact, the up quark has charge $Q = +\frac{2}{3}$ and the down quark has charge $Q = -\frac{1}{3}$, if the electron has electric charge $Q = -1$. These relations are readily comprehensible if leptons and quarks are composed of the same kinds of particles.

Secondly, the leptons and quarks may be grouped into three generations: (i) $(\nu_e, e^-, u, d)$, (ii) $(\nu_\mu, \mu^-, c, s)$ and (iii) $(\nu_\tau, \tau^-, t, b)$, with each generation containing particles which have similar properties. Corresponding to the electron, $e^-$, the second and third generations include the muon, $\mu^-$, and the tau particle, $\tau^-$, respectively. Each generation contains a neutrino associated with the corresponding leptons: the electron neutrino, $\nu_e$, the muon neutrino, $\nu_\mu$, and the tau neutrino, $\nu_\tau$. In addition, each generation contains a quark with $Q = +\frac{2}{3}$ (the $u$, $c$ and $t$ quarks) and a quark with $Q = -\frac{1}{3}$ (the $d$, $s$ and $b$ quarks). Each pair of leptons, e.g. $(\nu_e, e^-)$, and each pair of quarks, e.g. $(u, d)$, are connected by isospin symmetries, otherwise the grouping into the three families is according to increasing mass of the corresponding family members. The existence of three repeating patterns suggests strongly that the members of each generation are composites.

Thirdly, the GM, which provides a *unified* classification scheme for leptons and quarks, also indicates that these particles are intimately related. It has been demonstrated (Robson, 2004) that this unified classification scheme leads to a relation between strong isospin ($I$) and weak isospin ($i$) symmetries. In particular, their third components are related by an equation:

$$i_3 = I_3 + \frac{1}{2}g\,, \tag{24}$$

where $g$ is the generation quantum number. In addition, electric charge is related to $I_3$, $p$, $g$ and $i_3$ by the equations:

$$Q = I_3 + \frac{1}{2}(p + g) = i_3 + \frac{1}{2}p\,. \tag{25}$$

These relations are valid for both leptons and quarks and suggest that there exists an underlying flavor $SU(3)$ symmetry. The simplest conjecture is that this new flavor symmetry is connected with the substructure of leptons and quarks, analogous to the flavor $SU(3)$ symmetry underlying the quark structure of the lower mass hadrons in the Eightfold Way (Gell-Mann and Ne'eman, 1964).

The CGM description of the first generation is based upon the two-particle models of Harari (1979) and Shupe (1979), which are very similar and provide an economical and impressive description of the first generation of leptons and quarks. Both models treat leptons and quarks as composites of two kinds of spin-1/2 particles, which Harari named "rishons" from the Hebrew word for first or primary. This name has been adopted for the constituents of leptons and quarks. The CGM is constructed within the framework of the GM, i.e. the *same* kind of additive quantum numbers are assigned to the constituents of both leptons and quarks, as were previously allotted in the GM to leptons and quarks (see Table 2).

In the Harari-Shupe Model (HSM), two elementary spin-1/2 rishons and their corresponding antiparticles are employed to construct the leptons and quarks: (i) a $T$-rishon with $Q = +1/3$ and (ii) a $V$-rishon with $Q = 0$. Their antiparticles (denoted in the usual way by a bar over the defining particle symbol) are a $\bar{T}$-antirishon with $Q = -1/3$ and a $\bar{V}$-antirishon with $Q = 0$, respectively. Each spin-1/2 lepton and quark is composed of three rishons/antirishons.

Table 3 shows the proposed structures of the first generation of leptons and quarks in the HSM.

| particle | *structure* | $Q$ |
|---|---|---|
| $e^+$ | $TTT$ | $+1$ |
| $u$ | $TTV, TVT, VTT$ | $+\frac{2}{3}$ |
| $\bar{d}$ | $TVV, VTV, VVT$ | $+\frac{1}{3}$ |
| $\nu_e$ | $VVV$ | 0 |
| $\bar{\nu}_e$ | $\bar{V}\bar{V}\bar{V}$ | 0 |
| $d$ | $\bar{T}\bar{V}\bar{V}, \bar{V}\bar{T}\bar{V}, \bar{V}\bar{V}\bar{T}$ | $-\frac{1}{3}$ |
| $\bar{u}$ | $\bar{T}\bar{T}\bar{V}, \bar{T}\bar{V}\bar{T}, \bar{V}\bar{T}\bar{T}$ | $-\frac{2}{3}$ |
| $e^-$ | $\bar{T}\bar{T}\bar{T}$ | $-1$ |

Table 3. HSM of first generation of leptons and quarks

It should be noted that no composite particle involves mixtures of rishons and antirishons, as emphasized by Shupe. Both Harari and Shupe noted that quarks contained mixtures of the two kinds of rishons, whereas leptons did not. They concluded that the concept of color related to the different internal arrangements of the rishons in a quark: initially the ordering $TTV$, $TVT$ and $VTT$ was associated with the three colors of the $u$-quark. However, at this stage, no underlying mechanism was suggested for color. Later, a dynamical basis was proposed by Harari and Seiberg (1981), who were led to consider color-type local gauged $SU(3)$ symmetries, namely $SU(3)_C \times SU(3)_H$, at the rishon level. They proposed a new super-strong color-type (hypercolor) interaction corresponding to the $SU(3)_H$ symmetry, mediated by massless *hypergluons*, which is responsible for binding rishons together to form hypercolorless leptons or quarks. This interaction was assumed to be analogous to the strong color interaction of the SM, mediated by massless gluons, which is responsible for binding quarks together to form baryons or mesons. However, in this dynamical rishon model, the color force corresponding to the $SU(3)_C$ symmetry is also retained, with the $T$-rishons and $V$-rishons carrying colors and anticolors. respectively, so that leptons are colorless but quarks are colored. Similar proposals were made by others (Casalbuoni and Gatto, 1980; Squires, 1980; 1981). In each of these proposals, both the color force and the new hypercolor interaction are assumed to exist independently of one another so that the original rishon model loses some of its economical description. Furthermore, the HSM does not provide a satisfactory understanding of the second and third generations of leptons and quarks.

| rishon | $Q$ | $p$ | $g$ |
|---|---|---|---|
| $T$ | $+\frac{1}{3}$ | $+\frac{1}{3}$ | 0 |
| $V$ | 0 | $+\frac{1}{3}$ | 0 |
| $U$ | 0 | $+\frac{1}{3}$ | $-1$ |

Table 4. CGM additive quantum numbers for rishons

In order to overcome some of the deficiencies of the simple HSM, the two-rishon model was extended (Robson, 2005; 2011a), within the framework of the GM, in several ways.

Firstly, following the suggested existence of an $SU(3)$ flavor symmetry underlying the substructure of leptons and quarks by Eq. (25), a third type of rishon, the $U$-rishon, is introduced. This $U$-rishon has $Q = 0$ but carries a non-zero generation quantum number, $g = -1$ (both the $T$-rishon and the $V$-rishon are assumed to have $g = 0$). Thus, the CGM treats leptons and quarks as composites of *three* kinds of spin-1/2 rishons, although the $U$-rishon is only involved in the second and third generations.

Secondly, in the CGM, each rishon is allotted both a particle number $p$ and a generation quantum number $g$. Table 4 gives the three additive quantum numbers allotted to the three kinds of rishons. It should be noted that for each rishon additive quantum number $N$, the corresponding antirishon has the additive quantum number $-N$.

Historically, the term "particle" defines matter that is naturally occurring, especially electrons. In the CGM it is convenient to define a matter "particle" to have $p > 0$, with the antiparticle having $p < 0$. This definition of a matter particle leads to a modification of the HSM structures of the leptons and quarks which comprise the first generation. Essentially, the roles of the $V$-rishon and its antiparticle $\bar{V}$ are interchanged in the CGM compared with the HSM. Table 5 gives the CGM structures for the first generation of leptons and quarks. The particle number $p$ is clearly given by $\frac{1}{3}$(number of rishons - number of antirishons). Thus the $u$-quark has $p = +\frac{1}{3}$, since it contains two $T$-rishons and one $\bar{V}$-antirishon. It should be noted that it is essential for the $u$-quark to contain a $\bar{V}$-antirishon ($p = -\frac{1}{3}$) rather than a $V$-rishon ($p = +\frac{1}{3}$) to obtain a value of $p = +\frac{1}{3}$, corresponding to baryon number $A = +\frac{1}{3}$ in the SM.

In the CGM, no significance is attached to the ordering of the $T$-rishons and the $\bar{V}$-antirishons (compare HSM) so that, e.g. the structures $TT\bar{V}$,$T\bar{V}T$ and $\bar{V}TT$ for the $u$-quark are considered to be equivalent. The concept of color is treated differently in the CGM: it is assumed that all three rishons, $T$, $V$ and $U$ carry a color charge, red, green or blue, while their antiparticles carry an anticolor charge, antired, antigreen or antiblue. The CGM postulates a strong color-type interaction corresponding to a local gauged $SU(3)_C$ symmetry (analogous to QCD) and mediated by massless *hypergluons*, which is responsible for binding rishons and antirishons together to form colorless leptons and colored quarks. The proposed structures of the quarks requires the composite quarks to have a color charge so that the dominant residual interaction between quarks is essentially the same as that between rishons, and consequently the composite quarks behave very like the elementary quarks of the SM. In the CGM we retain the term "hypergluon" as the mediator of the strong color interaction, rather than the term "gluon" employed in the SM, because it is the rishons rather than the quarks, which carry an elementary color charge.

In the CGM each lepton of the first generation (Table 5) is assumed to be colorless, consisting of three rishons (or antirishons), each with a different color (or anticolor), analogous to the

| particle | *structure* | $Q$ | $p$ | $g$ |
|---|---|---|---|---|
| $e^+$ | $TTT$ | $+1$ | $+1$ | 0 |
| $u$ | $TT\bar{V}$ | $+\frac{2}{3}$ | $+\frac{1}{3}$ | 0 |
| $\bar{d}$ | $T\bar{V}\bar{V}$ | $+\frac{1}{3}$ | $-\frac{1}{3}$ | 0 |
| $\nu_e$ | $\bar{V}\bar{V}\bar{V}$ | 0 | $-1$ | 0 |
| $\bar{\nu}_e$ | $VVV$ | 0 | $+1$ | 0 |
| $d$ | $\bar{T}VV$ | $-\frac{1}{3}$ | $+\frac{1}{3}$ | 0 |
| $\bar{u}$ | $\bar{T}\bar{T}V$ | $-\frac{2}{3}$ | $-\frac{1}{3}$ | 0 |
| $e^-$ | $\bar{T}\bar{T}\bar{T}$ | $-1$ | $-1$ | 0 |

Table 5. CGM of first generation of leptons and quarks

baryons (or antibaryons) of the SM. These leptons are built out of $T$- and $V$-rishons or their antiparticles $\bar{T}$ and $\bar{V}$, all of which have generation quantum number $g = 0$.

It is envisaged that each lepton of the first generation exists in an antisymmetric three-particle color state, which physically assumes a quantum mechanical triangular distribution of the three differently colored identical rishons (or antirishons), since each of the three color interactions between pairs of rishons (or antirishons) is expected to be strongly attractive (Halzen and Martin, 1984).

In the CGM, it is assumed that each quark of the first generation is a composite of a colored rishon and a colorless rishon-antirishon pair, $(T\bar{V})$ or $(V\bar{T})$, so that the quarks carry a color charge. Similarly, the antiquarks are a composite of an anticolored antirishon and a colorless rishon-antirishon pair, so that the antiquarks carry an anticolor charge.

In order to preserve the universality of the CC weak interaction processes involving first generation quarks, e.g. the transition $d \to u + W^-$, it is assumed that the first generation quarks have the general color structures:

$$\text{up quark}: \quad T_C(T_{C'}\bar{V}_{\bar{C}'}), \quad \text{down quark}: \quad V_C(V_{C'}\bar{T}_{\bar{C}'}), \quad \text{with } C' \neq C. \tag{26}$$

Thus a red $u$-quark and a red $d$-quark have the general color structures:

$$u_r = T_r(T_g\bar{V}_{\bar{g}} + T_b\bar{V}_{\bar{b}})/\sqrt{2}, \tag{27}$$

and

$$d_r = V_r(V_g\bar{T}_{\bar{g}} + V_b\bar{T}_{\bar{b}})/\sqrt{2}, \tag{28}$$

respectively. For $d_r \to u_r + W^-$, conserving color, one has the two transitions:

$$V_rV_g\bar{T}_{\bar{g}} \to T_rT_b\bar{V}_{\bar{b}} + V_rV_gV_b\bar{T}_{\bar{r}}\bar{T}_{\bar{g}}\bar{T}_{\bar{b}} \tag{29}$$

and

$$V_rV_b\bar{T}_{\bar{b}} \to T_rT_g\bar{V}_{\bar{g}} + V_rV_gV_b\bar{T}_{\bar{r}}\bar{T}_{\bar{g}}\bar{T}_{\bar{b}}, \tag{30}$$

which take place with equal probabilities. In these transitions, the $W^-$ boson is assumed to be a three $\bar{T}$-antirishon and a three $V$-rishon colorless composite particle with additive quantum numbers $Q = -1, p = g = 0$. The corresponding $W^+$ boson has the structure $[T_rT_gT_b\bar{V}_{\bar{r}}\bar{V}_{\bar{g}}\bar{V}_{\bar{b}}]$,

| particle | structure | $Q$ | $p$ | $g$ |
|---|---|---|---|---|
| $\mu^+$ | $TTT\Pi$ | $+1$ | $+1$ | $\pm1$ |
| $c$ | $TT\bar{V}\Pi$ | $+\frac{2}{3}$ | $+\frac{1}{3}$ | $\pm1$ |
| $\bar{s}$ | $T\bar{V}\bar{V}\Pi$ | $+\frac{1}{3}$ | $-\frac{1}{3}$ | $\pm1$ |
| $\nu_\mu$ | $\bar{V}\bar{V}\bar{V}\Pi$ | $0$ | $-1$ | $\pm1$ |
| $\bar{\nu}_\mu$ | $VVV\Pi$ | $0$ | $+1$ | $\pm1$ |
| $s$ | $\bar{T}VV\Pi$ | $-\frac{1}{3}$ | $+\frac{1}{3}$ | $\pm1$ |
| $\bar{c}$ | $\bar{T}\bar{T}V\Pi$ | $-\frac{2}{3}$ | $-\frac{1}{3}$ | $\pm1$ |
| $\mu^-$ | $\bar{T}\bar{T}\bar{T}\Pi$ | $-1$ | $-1$ | $\pm1$ |

Table 6. CGM of second generation of leptons and quarks

consisting of a colorless set of three $T$-rishons and a colorless set of three $\bar{V}$-antirishons with additive quantum numbers $Q = +1$, $p = g = 0$ (Robson, 2005).

The rishon structures of the second generation particles are the same as the corresponding particles of the first generation plus the addition of a colorless rishon-antirishon pair, $\Pi$, where

$$\Pi = [(\bar{U}V) + (\bar{V}U)]/\sqrt{2}\,, \tag{31}$$

which is a quantum mechanical mixture of $(\bar{U}V)$ and $(\bar{V}U)$, which have $Q = p = 0$ but $g = \pm1$, respectively. In this way, the pattern for the first generation is repeated for the second generation. Table 6 gives the CGM structures for the second generation of leptons and quarks.

It should be noted that for any given transition the generation quantum number is required to be conserved, although each particle of the second generation has two possible values of $g$. For example, the decay

$$\mu^- \to \nu_\mu + W^-\,, \tag{32}$$

at the rishon level may be written

$$\bar{T}\bar{T}\bar{T}\Pi \to \bar{V}\bar{V}\bar{V}\Pi + \bar{T}\bar{T}\bar{T}VVV\,, \tag{33}$$

which proceeds via the two transitions:

$$\bar{T}\bar{T}\bar{T}(\bar{U}V) \to \bar{V}\bar{V}\bar{V}(\bar{U}V) + \bar{T}\bar{T}\bar{T}VVV \tag{34}$$

and

$$\bar{T}\bar{T}\bar{T}(\bar{V}U) \to \bar{V}\bar{V}\bar{V}(\bar{V}U) + \bar{T}\bar{T}\bar{T}VVV\,, \tag{35}$$

which take place with equal probabilities. In each case, the additional colorless rishon-antirishon pair, $(\bar{U}V)$ or $(\bar{V}U)$, essentially acts as a spectator during the CC weak interaction process.

The rishon structures of the third generation particles are the same as the corresponding particles of the first generation plus the addition of two rishon-antirishon pairs, which are a quantum mechanical mixture of $(\bar{U}V)$ and $(\bar{V}U)$ and, as for the second generation, are assumed to be colorless and have $Q = p = 0$ but $g = \pm1$. In this way the pattern of the first and second generation is continued for the third generation. Table 7 gives the CGM structures for the third generation of leptons and quarks.

| particle | *structure* | $Q$ | $p$ | $g$ |
|---|---|---|---|---|
| $\tau^+$ | $TTT\Pi\Pi\Pi$ | $+1$ | $+1$ | $0, \pm 2$ |
| $t$ | $TT\bar{V}\Pi\Pi\Pi$ | $+\frac{2}{3}$ | $+\frac{1}{3}$ | $0, \pm 2$ |
| $\bar{b}$ | $T\bar{V}\bar{V}\Pi\Pi\Pi$ | $+\frac{1}{3}$ | $-\frac{1}{3}$ | $0, \pm 2$ |
| $\nu_\tau$ | $\bar{V}\bar{V}\bar{V}\Pi\Pi\Pi$ | $0$ | $-1$ | $0, \pm 2$ |
| $\bar{\nu}_\tau$ | $VVV\Pi\Pi\Pi$ | $0$ | $+1$ | $0, \pm 2$ |
| $b$ | $\bar{T}VV\Pi\Pi\Pi$ | $-\frac{1}{3}$ | $+\frac{1}{3}$ | $0, \pm 2$ |
| $\bar{t}$ | $\bar{T}\bar{T}V\Pi\Pi\Pi$ | $-\frac{2}{3}$ | $-\frac{1}{3}$ | $0, \pm 2$ |
| $\tau^-$ | $\bar{T}\bar{T}\bar{T}\Pi\Pi\Pi$ | $-1$ | $-1$ | $0, \pm 2$ |

Table 7. CGM of third generation of leptons and quarks

The rishon structure of the $\tau^+$ particle is

$$TTT\Pi\Pi\Pi = TTT[(\bar{U}V)(\bar{U}V) + (\bar{U}V)(\bar{V}U) + (\bar{V}U)(\bar{U}V) + (\bar{V}U)(\bar{V}U)]/2 \tag{36}$$

and each particle of the third generation is a similar quantum mechanical mixture of $g = 0, \pm 2$ components. The color structures of both second and third generation leptons and quarks have been chosen so that the CC weak interactions are universal. In each case, the additional colorless rishon-antirishon pairs, $(\bar{U}V)$ and/or $(\bar{V}U)$, essentially act as spectators during any CC weak interaction process. Again it should be noted that for any given transition the generation quantum number is required to be conserved, although each particle of the third generation now has three possible values of $g$. Furthermore, in the CGM the three independent additive quantum numbers, charge $Q$, particle number $p$ and generation quantum number $g$, which are conserved in all interactions, correspond to the conservation of each of the three kinds of rishons (Robson, 2005):

$$n(T) - n(\bar{T}) = 3Q\,, \tag{37}$$

$$n(\bar{U}) - n(U) = g\,, \tag{38}$$

$$n(T) + n(V) + n(U) - n(\bar{T}) - n(\bar{V}) - n(\bar{U}) = 3p\,, \tag{39}$$

where $n(R)$ and $n(\bar{R})$ are the numbers of rishons and antirishons, respectively. Thus, the conservation of $g$ in weak interactions is a consequence of the conservation of the three kinds of rishons ($T$, $V$ and $U$), which also prohibits transitions between the third generation and the first generation via weak interactions even for $g = 0$ components of third generation particles.

### 3.4 Fundamental interactions of the GM

The GM recognizes only two fundamental interactions in nature: (i) the usual electromagnetic interaction and (ii) a strong color-type interaction, mediated by massless hypergluons, acting between color charged rishons and/or antirishons.

The only essential difference between the strong color interactions of the GM and the SM is that the former acts between color charged rishons and/or antirishons while the latter acts between color charged elementary quarks and/or antiquarks. For historical reasons we use the term "hypergluons" for the mediators of the strong color interactions at the rishon level, rather than the term "gluons" as employed in the SM, although the effective color interaction between composite quarks and/or composite antiquarks is very similar to that between the elementary quarks and/or elementary antiquarks of the SM.

In the GM both gravity and the weak interactions are considered to be residual interactions of the strong color interactions. Gravity will be discussed in some detail in Subsection 4.3. In the GM the weak interactions are assumed to be mediated by composite massive vector bosons, consisting of colorless sets of three rishons and three antirishons as discussed in the previous subsection, so that they are not elementary particles, associated with a $U(1) \times SU(2)_L$ local gauge theory as in the SM. The weak interactions are simply residual interactions of the CGM strong color force, which binds rishons and antirishons together, analogous to the strong nuclear interactions, mediated by massive mesons, being residual interactions of the strong color force of the SM, which binds quarks and antiquarks together. Since the weak interactions are not considered to be fundamental interactions arising from a local gauge theory, there is no requirement for the existence of a Higgs field to generate the boson masses within the framework of the GM (Robson, 2008).

## 4. Consequences

In this section it will be shown that new paradigms arising from the GM provide some understanding concerning: (i) the origin of mass; (ii) the mass hierarchy of leptons and quarks; (iii) the origin of gravity and (iv) the origin of "apparent" CP violation in the $K^0 - \bar{K}^0$ system.

### 4.1 Origin of mass

Einstein (1905) concluded that the mass of a body $m$ is a measure of its energy content $E$ and is given by

$$m = E/c^2 , \tag{40}$$

where $c$ is the speed of light in a vacuum. This relationship was first tested by Cockcroft and Walton (1932) using the nuclear transformation

$$^7\text{Li} + p \to 2\alpha + 17.2\text{ MeV} , \tag{41}$$

and it was found that the decrease in mass in this disintegration process was consistent with the observed release of energy, according to Eq. (40). Recently, relation (40) has been verified (Rainville *et al.*, 2005) to within 0.00004%, using very accurate measurements of the atomic-mass difference, $\Delta m$, and the corresponding $\gamma$-ray wavelength to determine $E$, the nuclear binding energy, for isotopes of silicon and sulfur.

It has been emphasized by Wilczek (2005) that approximate QCD calculations (Butler *et al.*, 1993; Aoki *et al.*, 2000; Davies *et al.*, 2004) obtain the observed masses of the neutron, proton and other baryons to an accuracy of within 10%. In these calculations, the assumed constituents, quarks and gluons, are taken to be massless. Wilczek concludes that the calculated masses of the hadrons arise from both the energy stored in the motion of the quarks and the energy of the gluon fields, according to Eq. (40): basically the mass of a hadron arises from internal energy.

Wilzcek (2005) has also discussed the underlying principles giving rise to the internal energy, hence the mass, of a hadron. The nature of the gluon color fields is such that they lead to a runaway growth of the fields surrounding an isolated color charge. In fact all this structure (via virtual gluons) implies that an isolated quark would have an infinite energy associated with it. This is the reason why isolated quarks are not seen. Nature requires these infinities to be essentially cancelled or at least made finite. It does this for hadrons in two ways: either by bringing an antiquark close to a quark (i.e forming a meson) or by bringing three quarks, one of each color, together (i.e. forming a baryon) so that in each case the composite hadron is colorless. However, quantum mechanics prevents the quark and the antiquark of opposite colors or the three quarks of different colors from being placed exactly at the same place. This means that the color fields are not exactly cancelled, although sufficiently it seems to remove the infinities associated with isolated quarks. The distribution of the quark-antiquark pairs or the system of three quarks is described by quantum mechanical wave functions. Many different patterns, corresponding to the various hadrons, occur. Each pattern has a characteristic energy, because the color fields are not entirely cancelled and because the quarks are somewhat localized. This characteristic energy, $E$, gives the characteristic mass, via Eq. (40), of the hadron.

The above picture, within the framework of the SM, provides an understanding of hadron masses as arising mainly from internal energies associated with the strong color interactions. However, as discussed in Subsection 2.2.3, the masses of the elementary particles of the SM, the leptons, the quarks and the $W$ and $Z$ bosons, are interpreted in a completely different way. A "condensate" called the Higgs scalar field (Englert and Brout, 1964; Higgs, 1964), analogous to the Cooper pairs in a superconducting material, is assumed to exist. This field couples, with an appropriate strength, to each lepton, quark and vector boson and endows an originally massless particle with its physical mass. Thus, the assumption of a Higgs field within the framework of the SM not only adds an extra field but also leads to the introduction of 14 new parameters. Moreover, as pointed out by Lyre (2008), the introduction of the Higgs field in the SM to spontaneously break the $U(1) \times SU(2)_L$ local gauge symmetry of the electroweak interaction to generate the masses of the $W$ and $Z$ bosons, simply corresponds mathematically to putting in "by hand" the masses of the elementary particles of the SM: the so-called Higgs mechanism does *not* provide any physical explanation for the origin of the masses of the leptons, quarks and the $W$ and $Z$ bosons.

In the CGM (Robson, 2005; 2011a), the elementary particles of the SM have a substructure, consisting of massless rishons and/or antirishons bound together by strong color interactions, mediated by massless neutral hypergluons. This model is very similar to that of the SM in which the quarks and/or antiquarks are bound together by strong color interactions, mediated by massless neutral gluons, to form hadrons. Since, as discussed above, the mass of a hadron arises mainly from the energy of its constituents, the CGM suggests (Robson, 2009) that the mass of a lepton, quark or vector boson arises entirely from the energy stored in the motion of its constituent rishons and/or antirishons and the energy of the color hypergluon fields, $E$, according to Eq. (40). A corollary of this idea is: *if a particle has mass, then it is composite*. Thus, unlike the SM, the GM provides a *unified* description of the origin of *all* mass.

### 4.2 Mass hierarchy of leptons and quarks

Table 8 shows the observed masses of the charged leptons together with the estimated masses of the quarks: the masses of the neutral leptons have not yet been determined but are known to be very small. Although the mass of a single quark is a somewhat abstract idea, since quarks do not exist as particles independent of the environment around them, the masses of the quarks may be inferred from mass differences between hadrons of similar composition. The strong binding within hadrons complicates the issue to some extent but rough estimates of the quark masses have been made (Veltman, 2003), which are sufficient for our purposes.

| Charge | 0 | $-1$ | $+\frac{2}{3}$ | $-\frac{1}{3}$ |
|---|---|---|---|---|
| Generation 1 | $\nu_e$ | $e^-$ | $u$ | $d$ |
| Mass | $<0.3$ eV | 0.511 MeV | 5 MeV | 10 MeV |
| Generation 2 | $\nu_\mu$ | $\mu^-$ | $c$ | $s$ |
| Mass | $<0.3$ eV | 106 MeV | 1.3 GeV | 200 MeV |
| Generation 3 | $\nu_\tau$ | $\tau^-$ | $t$ | $b$ |
| Mass | $<0.3$ eV | 1.78 GeV | 175 GeV | 4.5 GeV |

Table 8. Masses of leptons and quarks

The SM is unable to provide any understanding of either the existence of the three generations of leptons and quarks or their mass hierarchy indicated in Table 8; whereas the CGM suggests that both the existence and mass hierarchy of these three generations arise from the substructures of the leptons and quarks (Robson, 2009; 2011a).

Subsection 3.3 describes the proposed rishon and/or antirishon substructures of the three generations of leptons and quarks and indicates how the pattern of the first generation is followed by the second and third generations. Section 4.1 discusses the origin of mass in composite particles and postulates that the mass of a lepton or quark arises from the energy of its constituents.

In the CGM it is envisaged that the rishons and/or antirishons of each lepton or quark are very strongly localized, since to date there is no direct evidence for any substructure of these particles. Thus the constituents are expected to be distributed according to quantum mechanical wave functions, for which the product wave function is significant for only an *extremely small* volume of space so that the corresponding color fields are *almost cancelled*. The constituents of each lepton or quark are localized within a very small volume of space by strong color interactions acting between the colored rishons and/or antirishons. We call these *intra-fermion* color interactions. However, between any two leptons and/or quarks there will be a residual interaction, arising from the color interactions acting between the constituents of one fermion and the constituents of the other fermion. We refer to these interactions as *inter-fermion* color interactions. These will be associated with the gravitational interaction and are discussed in the next subsection.

The mass of each lepton or quark corresponds to a characteristic energy primarily associated with the intra-fermion color interactions. It is expected that the mass of a composite particle will be greater if the degree of localization of its constituents is smaller (i.e. the constituents are on average more widely separated). This is a consequence of the nature of the strong color interactions, which are assumed to possess the property of "asymptotic freedom" (Gross and Wilczek, 1973; Politzer, 1973), whereby the color interactions become stronger for larger separations of the color charges. In addition, it should be noted that the electromagnetic interactions between charged $T$-rishons or between charged $\bar{T}$-antirishons will also cause the degree of localization of the constituents to be smaller causing an increase in mass.

There is some evidence for the above expectations. The electron consists of three $\bar{T}$-antirishons, while the electron neutrino consists of three neutral $\bar{V}$-antirishons. Neglecting the electric charge carried by the $\bar{T}$-antirishon, it is expected that the electron and its neutrino would have identical masses, arising from the similar intra-fermion color interactions. However, it is anticipated that the electromagnetic interaction in the electron case will cause the $\bar{T}$-antirishons to be less localized than the $\bar{V}$-antirishons constituting the electron neutrino so that the electron will have a substantially greater characteristic energy and hence a greater mass than the electron neutrino, as observed. This large difference in the masses of the $e^-$ and $\nu_e$ leptons (see Table 8) indicates that the mass of a particle is extremely sensitive to the degree of localization of its constituents. Similarly, the up, charmed and top quarks, each containing two charged $T$-rishons, are expected to have a greater mass than their weak isospin partners, the down, strange and bottom quark, respectively, which contain only a single charged $\bar{T}$-antirishon. This is true provided one takes into account quark mixing (Evans and Robson, 2006) in the case of the up and down quarks, although Table 8 indicates that the down quark is more massive than the up quark, leading to the neutron having a greater mass than the proton. This is understood within the framework of the GM since due to the manner in which quark masses are estimated, it is the *weak eigenstate* quarks, whose masses are given in Table 8. Since each succeeding generation is significantly more massive than the previous one, any mixing will noticeably increase the mass of a lower generation quark. Thus the weak eigenstate $d'$-quark, which contains about 5% of the mass eigenstate $s$-quark, is expected to be significantly more massive than the mass eigenstate $d$-quark (see Subsection 3.2). We shall now discuss the mass hierarchy of the three generations of leptons and quarks in more detail.

It is envisaged that each lepton of the *first* generation exists in an antisymmetric three-particle color state, which physically assumes a quantum mechanical triangular distribution of the three differently colored identical rishons (or antirishons) since each of the three color interactions between pairs of rishons (or antirishons) is expected to be strongly attractive (Halzen and Martin, 1984). As indicated above, the charged leptons are predicted to have larger masses than the neutral leptons, since the electromagnetic interaction in the charged leptons will cause their constituent rishons (or antirishons) to be less localized than those constituting the uncharged leptons, leading to a substantially greater characteristic energy and a correspondingly greater mass.

In the CGM, each quark of the *first* generation is a composite of a colored rishon and a colorless rishon-antirishon pair, $(T\bar{V})$ or a $(V\bar{T})$ (see Table 5). This color charge structure of the quarks is expected to lead to a quantum mechanical linear distribution of the constituent rishons and antirishons, corresponding to a considerably larger mass than that of the leptons, since the constituents of the quarks are less localized. This is a consequence of the character (i.e. attractive or repulsive) of the color interactions at small distances (Halzen and Martin, 1984). The general rules for small distances of separation are:

(i) rishons (or antirishons) of like colors (or anticolors) repel: those having different colors (or anticolors) attract, unless their colors (or anticolors) are interchanged and the two rishons (or antirishons) do not exist in an antisymmetric color state (e.g. as in the case of leptons);

(ii) rishons and antirishons of opposite colors attract but otherwise repel.

Furthermore, the electromagnetic interaction occurring within the up quark, leads one to expect it to have a larger mass than that of the down quark.

Each lepton of the *second* generation is envisaged to basically exist in an antisymmetric three-particle color state, which physically assumes a quantum mechanical triangular distribution of the three differently colored identical rishons (or antirishons), as for the corresponding lepton of the first generation. The additional colorless rishon-antirishon pair, $(V\bar{U})$ or $(U\bar{V})$, is expected to be attached externally to this triangular distribution, leading quantum mechanically to a less localized distribution of the constituent rishons and/or antirishons, so that the lepton has a significantly larger mass than its corresponding first generation lepton.

Each quark of the *second* generation has a similar structure to that of the corresponding quark of the first generation, with the additional colorless rishon-antirishon pair, $(V\bar{U})$ or $(U\bar{V})$, attached quantum mechanically so that the whole rishon structure is essentially a linear distribution of the constituent rishons and antirishons. This structure is expected to be less localized, leading to a larger mass relative to that of the corresponding quark of the first generation, with the charmed quark having a greater mass than the strange quark, arising from the electromagnetic repulsion of its constituent two charged $T$-rishons.

Each lepton of the *third* generation is considered to basically exist in an antisymmetric three-particle color state, which physically assumes a quantum mechanical triangular distribution of the three differently colored identical rishons (or antirishons), as for the corresponding leptons of the first and second generations. The two additional colorless rishon-antirishon pairs, $(V\bar{U})(V\bar{U})$, $(V\bar{U})(U\bar{V})$ or $(U\bar{V})(U\bar{V})$, are expected to be attached externally to this triangular distribution, leading to a considerably less localized quantum mechanical distribution of the constituent rishons and/or antirishons, so that the lepton has a significantly larger mass than its corresponding second generation lepton.

Each quark of the *third* generation has a similar structure to that of the first generation, with the additional two rishon-antirishon pairs $(V\bar{U})$ and/or $(U\bar{V})$ attached quantum mechanically so that the whole rishon structure is essentially a linear distribution of the constituent rishons and antirishons. This structure is expected to be even less localized, leading to a larger mass relative to that of the corresponding quark of the second generation, with the top quark having a greater mass than the bottom quark, arising from the electromagnetic repulsion of its constituent two charged $T$-rishons.

The above is a qualitative description of the mass hierarchy of the three generations of leptons and quarks, based on the degree of localization of their constituent rishons and/or antirishons. However, in principle, it should be possible to calculate the actual masses of the leptons and quarks by carrying out QCD-type computations, analogous to those employed for determining the masses of the proton and other baryons within the framework of the SM (Butler *et al.*, 1993; Aoki *et al.*, 2000; Davies *et al.*, 2004).

### 4.3 Origin of gravity

Robson (2009) proposed that the residual interaction, arising from the incomplete cancellation of the inter-fermion color interactions acting between the rishons and/or antirishons of one colorless particle and those of another colorless particle, may be identified with the usual gravitational interaction, since it has several properties associated with that interaction: universality, infinite range and very weak strength. Based upon this earlier conjecture, Robson (2011a) has presented a quantum theory of gravity, described below, leading approximately to Newton's law of universal gravitation.

The mass of a body of ordinary matter is essentially the total mass of its constituent electrons, protons and neutrons. It should be noted that these masses will depend upon the environment in which the particle exists: e.g. the mass of a proton in an atom of helium will differ slightly from that of a proton in an atom of lead. In the CGM, each of these three particles is considered to be colorless. The electron is composed of three $\bar{T}$-antirishons, each carrying a different anticolor charge, antired, antigreen or antiblue. Both the proton and neutron are envisaged (as in the SM) to be composed of three quarks, each carrying a different color charge, red, green or blue. All three particles are assumed to be essentially in a three-color antisymmetric state, so that their behavior with respect to the strong color interactions is expected basically to be the same. This similar behavior suggests that the proposed residual interaction has several properties associated with the usual gravitational interaction.

Firstly, the residual interaction between any two of the above colorless particles, arising from the inter-fermion color interactions, is predicted to be of a *universal* character.

Secondly, assuming that the strong color fields are almost completely cancelled at large distances, it seems plausible that the residual interaction, mediated by massless hypergluons, should have an infinite range, and tend to zero as $1/r^2$. These properties may be attributed to the fact that the constituents of each colorless particle are very strongly localized so that the strength of the residual interaction is *extremely weak*, and consequently the hypergluon self-interactions are also practically negligible. This means that one may consider the color interactions using a perturbation approach: the residual color interaction is the sum of all the two-particle color charge interactions, each of which may be treated perturbatively, i.e. as a single hypergluon exchange. Using the color factors (Halzen and Martin, 1984) appropriate for the $SU(3)$ gauge field, one finds that the residual color interactions between any two colorless particles (electron, neutron or proton) are each *attractive*.

Since the mass of a body of ordinary matter is essentially the total mass of its constituent electrons, neutrons and protons, the total interaction between two bodies of masses, $m_1$ and $m_2$, will be the sum of all the two-particle contributions so that the total interaction will be proportional to the product of these two masses, $m_1 m_2$, provided that each two-particle interaction contribution is also proportional to the product of the masses of the two particles.

This latter requirement may be understood if each electron, neutron or proton is considered physically to be essentially a quantum mechanical triangular distribution of three differently colored rishons or antirishons. In this case, each particle may be viewed as a distribution of three color charges throughout a small volume of space with each color charge having a certain probability of being at a particular point, determined by its corresponding color wave function. The total residual interaction between two colorless particles will then be the sum of all the intrinsic interactions acting between a particular triangular distribution of one particle with that of the other particle.

Now the mass $m$ of each colorless particle is considered to be given by $m = E/c^2$, where $E$ is a characteristic energy, determined by the degree of localization of its constituent rishons and/or antirishons. Thus the significant volume of space occupied by the triangular distribution of the three differently colored rishons or antirishons is larger the greater the mass of the particle. Moreover, due to antiscreening effects (Gross and Wilczek, 1973; Politzer, 1973) of the strong color fields, the average strength of the color charge within each unit volume of the larger localized volume of space will be increased. If one assumes that the mass of a particle is proportional to the integrated sum of the intra-fermion interactions within the significant volume of space occupied by the triangular distribution, then the total residual interaction between two such colorless particles will be proportional to the product of their masses.

Thus the residual color interaction between two colorless bodies of masses, $m_1$ and $m_2$, is proportional to the product of these masses and moreover is expected to depend *approximately* as the inverse square of their distance of separation $r$, i.e. as $1/r^2$, in accordance with Newton's law of universal gravitation. The approximate dependence on the inverse square law is expected to arise from the effect of hypergluon self-interactions, especially for large separations. Such deviations from an inverse square law do not occur for electromagnetic interactions, since there are no corresponding photon self-interactions.

### 4.4 Mixed-quark states in hadrons

As discussed in Subsection 3.2 the GM postulates that hadrons are composed of weak eigenstate quarks rather than mass eigenstate quarks as in the SM. This gives rise to several important consequences (Evans and Robson, 2006; Morrison and Robson, 2009; Robson, 2011b; 2011c).

Firstly, hadrons composed of mixed-quark states might seem to suggest that the electromagnetic and strong interaction processes between mass eigenstate hadron components are not consistent with the fact that weak interaction processes occur between weak eigenstate quarks. However, since the electromagnetic and strong interactions are flavor independent: the down, strange and bottom quarks carry the same electric and color charges so that the weak eigenstate quarks have the same magnitude of electric and color charge as the mass eigenstate quarks. Consequently, the weak interaction is the *only* interaction in which the quark-mixing phenomenon can be detected.

Secondly, the occurrence of mixed-quark states in hadrons implies the existence of higher generation quarks in hadrons. In particular, the GM predicts that the proton contains $\approx 1.7\%$ of strange quarks, while the neutron having two $d'$-quarks contains $\approx 3.4\%$ of strange quarks. Recent experiments (Maas *et al.*, 2005; Armstrong *et al*, 2005) have provided some evidence for the existence of strange quarks in the proton. However, to date the experimental data are compatible with the predictions of both the GM and the SM ($\ll 1.7\%$).

Thirdly, the presence of strange quarks in nucleons explains why the mass of the neutron is greater than the mass of a proton, so that the proton is stable. This arises because the mass of the weak eigenstate $d'$-quark is larger than the mass of the $u$-quark, although the mass eigenstate $d$-quark is expected to be smaller than that of the $u$-quark, as discussed in the previous section.

Another consequence of the presence of mixed-quark states in hadrons is that mixed-quark states may have mixed parity. In the CGM the constituents of quarks are rishons and/or antirishons. If one assumes the simple convention that all rishons have positive parity and all their antiparticles have negative parity, one finds that the down and strange quarks have opposite intrinsic parities, according to the proposed structures of these quarks in the CGM: the $d$-quark (see Table 5) consists of two rishons and one antirishon ($P_d = -1$), while the $s$-quark (see Table 6) consists of three rishons and two antirishons ($P_s = +1$). The $u$-quark consists of two rishons and one antirishon so that $P_u = -1$, and the antiparicles of these three quarks have the corresponding opposite parities: $P_{\bar{d}} = +1$, $P_{\bar{s}} = -1$ and $P_{\bar{u}} = +1$.

In the SM the intrinsic parity of the charged pions is assumed to be $P_\pi = -1$. This result was established by Chinowsky and Steinberger (1954), using the capture of negatively charged pions in deuterium to form two neutrons, and led to the overthrow of the conservation of both parity (P) and charge-conjugation (C) (Lee and Yang, 1956; Wu *et al.*, 1957; Garwin *et al.*, 1957; Friedman and Telegdi, 1957) and later combined CP conservation (Christenson *et al.*, 1964). Recently, Robson (2011b) has demonstrated that this experiment is also compatible with the mixed-parity nature of the $\pi^-$ predicted by the CGM: $\approx (0.95P_d + 0.05P_s)$, with $P_d = -1$ and $P_s = +1$. This implies that the original determination of the parity of the negatively charged pion is *not* conclusive, if the pion has a complex substructure as in the CGM. Similarly, Robson (2011c) has shown that the recent determination (Abouzaid *et al.*, 2008) of the parity of the neutral pion, using the double Dalitz decay $\pi^0 \to e^+e^-e^+e^-$ is also compatible with the mixed-parity nature of the neutral pion predicted by the CGM.

This new concept of mixed-parity states in hadrons, based upon the existence of weak eigenstate quarks in hadrons and the composite nature of the mass eigenstate quarks, leads to an understanding of CP symmetry in nature. This is discussed in the following subsection.

### 4.5 CP violation in the $K^0 - \bar{K}^0$ system

Gell-Mann and Pais (1955) considered the behavior of neutral particles under the charge-conjugation operator C. In particular they considered the $K^0$ meson and realized that unlike the photon and the neutral pion, which transform into themselves under the C operator so that they are their own antiparticles, the antiparticle of the $K^0$ meson (strangeness $S = +1$), $\bar{K}^0$, was a distinct particle, since it had a different strangeness quantum number ($S = -1$). They concluded that the two neutral mesons, $K^0$ and $\bar{K}^0$, are degenerate particles that exhibit unusual properties, since they can transform into each other via weak interactions such as

$$K^0 \rightleftharpoons \pi^+\pi^- \rightleftharpoons \bar{K}^0. \tag{42}$$

In order to treat this novel situation, Gell-Mann and Pais suggested that it was more convenient to employ different particle states, rather than $K^0$ and $\bar{K}^0$, to describe neutral kaon decay. They suggested the following representative states:

$$K_1^0 = (K^0 + \bar{K}^0)/\sqrt{2}, \quad K_2^0 = (K^0 - \bar{K}^0)/\sqrt{2}, \tag{43}$$

and concluded that these particle states must have different decay modes and lifetimes. In particular they concluded that $K_1^0$ could decay to two charged pions, while $K_2^0$ would have a longer lifetime and more complex decay modes. This conclusion was based upon the conservation of C in the weak interaction processes: both $K_1^0$ and the $\pi^+\pi^-$ system are even (i.e. $C = +1$) under the C operation.

The particle-mixing theory of Gell-Mann and Pais was confirmed in 1957 by experiment, in spite of the incorrect assumption of C invariance in weak interaction processes. Following the discovery in 1957 of both C and P violation in weak interaction processes, the particle-mixing theory led to a suggestion by Landau (1957) that the weak interactions may be invariant under the combined operation CP.

Landau's suggestion implied that the Gell-Mann–Pais model of neutral kaons would still apply if the states, $K_1^0$ and $K_2^0$, were eigenstates of CP with eigenvalues $+1$ and $-1$, respectively. Since the charged pions were considered to have intrinsic parity $P_\pi = -1$, it was clear that only the $K_1^0$ state could decay to two charged pions, if CP was conserved.

The suggestion of Landau was accepted for several years since it nicely restored some degree of symmetry in weak interaction processes. However, the surprising discovery (Christenson *et al.*, 1964) of the decay of the long-lived neutral $K^0$ meson to two charged pions led to the conclusion that CP is violated in the weak interaction. The observed violation of CP conservation turned out to be very small ($\approx 0.2\%$) compared with the maximal violations ($\approx 100\%$) of both P and C conservation separately. Indeed the very smallness of the apparent CP violation led to a variety of suggestions explaining it in a CP-conserving way (Kabir, 1968; Franklin, 1986). However, these efforts were unsuccessful and CP violation in weak interactions was accepted.

An immediate consequence of this was that the role of $K_1^0$ (CP = +1) and $K_2^0$ (CP = −1), defined in Eqs. (43), was replaced by two new particle states, corresponding to the short-lived ($K_S^0$) and long-lived ($K_L^0$) neutral kaons:

$$K_S^0 = (K_1^0 + \epsilon K_2^0)/(1+|\epsilon|^2)^{\frac{1}{2}} \,, \quad K_L^0 = (K_2^0 + \epsilon K_1^0)/(1+|\epsilon|^2)^{\frac{1}{2}} \,, \tag{44}$$

where the small complex parameter $\epsilon$ is a measure of the CP impurity in the eigenstates $K_S^0$ and $K_L^0$. This method of describing CP violation in the Standard Model (SM), by introducing mixing of CP eigenstates, is called 'indirect CP violation'. It is essentially a phenomenological approach with the parameter $\epsilon$ to be determined by experiment.

Another method of introducing CP violation into the SM was proposed by Kobayashi and Maskawa (1973). By extending the idea of 'Cabibbo mixing' (see Subsection 2.2.3) to three generations, they demonstrated that this allowed a complex phase to be introduced into the quark-mixing (CKM) matrix, permitting CP violation to be directly incorporated into the weak interaction. This phenomenological method has within the framework of the SM successfully accounted for both the indirect CP violation discovered by Christenson *et al.* in 1964 and the "direct CP violation" related to the decay processes of the neutral kaons (Kleinknecht, 2003). However, to date, the phenomenological approach has not been able to provide an *a priori* reason for CP violation to occur nor to indicate the magnitude of any such violation.

Recently, Morrison and Robson (2009) have demonstrated that the indirect CP violation observed by Christenson *et al.* (1964) can be described in terms of mixed-quark states in hadrons. In addition, the rate of the decay of the $K_L^0$ meson relative to the decay into all charged modes is estimated accurately in terms of the Cabibbo-mixing angle.

In the CGM the $K^0$ and $\bar{K}^0$ mesons have the weak eigenstate quark structures $[d'\bar{s}']$ and $[s'\bar{d}']$, respectively. Neglecting the very small mixing components arising from the third generation, Morrison and Robson show that the long-lived neutral kaon, $K_L^0$, exists in a CP = -1 eigenstate as in the SM. On the other hand, the charged $2\pi$ system:

$$\begin{aligned}\pi^+\pi^- &= [u\bar{d}'][d'\bar{u}] \\ &= [u\bar{d}][d\bar{u}]\cos^2\theta_c + [u\bar{s}][s\bar{u}]\sin^2\theta_c + [u\bar{s}][d\bar{u}]\sin\theta_c\cos\theta_c \\ &\quad +[u\bar{d}][s\bar{u}])\sin\theta_c\cos\theta_c \,.\end{aligned} \tag{45}$$

For the assumed parities (see Subsection 4.4) of the quarks and antiquarks involved in Eq. (45), it is seen that the first two components are eigenstates of CP = +1, while the remaining two components $[u\bar{s}][d\bar{u}]$ and $[u\bar{d}][s\bar{u}]$, with amplitude $\sin\theta_c\cos\theta_c$ are not individually eigenstates of CP. However, taken together, the state ($[u\bar{s}][d\bar{u}]$ + $[u\bar{d}][s\bar{u}]$) is an eigenstate of CP with eigenvalue CP = -1. Taking the square of the product of the amplitudes of the two components comprising the CP = -1 eigenstate to be the "joint probability" of those two states existing together simultaneously, one can calculate that this probability is given by $(\sin\theta_c \cos\theta_c)^4$ = $2.34 \times 10^{-3}$, using $\cos\theta_c = 0.9742$ (Amsler *et al.*, 2008). Thus, the existence of a small component of the $\pi^+\pi^-$ system with eigenvalue CP = -1 indicates that the $K_L^0$ meson can decay to the charged $2\pi$ system without violating CP conservation. Moreover, the estimated decay rate is in good agreement with experimental data (Amsler *et al.*, 2008).

## 5. Summary and future prospects

The GM, which contains fewer elementary particles (27 counting both particles and antiparticles and their three different color forms) and only two fundamental interactions (the electromagnetic and strong color interactions), has been presented as a viable simpler alternative to the SM (61 elementary particles and four fundamental interactions).

In addition, the GM has provided new paradigms for particle physics, which have led to a new understanding of several phenomena not addressed by the SM. In particular, (i) the mass of a particle is attributed to the energy content of its constituents so that there is no requirement for the Higgs mechanism; (ii) the mass hierarchy of the three generations of leptons and quarks is described by the degree of localization of their constituent rishons and/or antirishons; (iii) gravity is interpreted as a quantum mechanical residual interaction of the strong color interaction, which binds rishons and/or antirishons together to form all kinds of matter and (iv) the decay of the long-lived neutral kaon is understood in terms of mixed-quark states in hadrons and not CP violation.

The GM also predicts that the mass of a free neutron is greater than the mass of a free proton so that the free proton is stable. In addition, the model predicts the existence of higher generation quarks in hadrons, which in turn predicts mixed-parity states in hadrons. Further experimentation is required to verify these predictions and thereby strengthen the Generation Model.

## 6. References

Abouzaid, E. *et al.* (2008), Determination of the Parity of the Neutral Pion via its Four-Electron Decay, *Physical Review Letters*, Vol. 100, No. 18, 182001 (5 pages).

Aitchison I.J.R. and Hey, A.J.G. (1982), *Gauge Theories in Particle Physics* (Adam Hilger Ltd, Bristol).

Amsler, C. *et al.* (2008), Summary Tables of Particle Properties, *Physics Letters B*, Vol. 667, Nos. 1-5, pp. 31-100.

Aoki, S. *et al.* (2000), Quenched Light Hadron Spectrum, *Physical Review Letters*, Vol. 84, No. 2, pp. 238-241.

Armstrong, D.S. *et al.* (2005), Strange-Quark Contributions to Parity-Violating Asymmetries in the Forward G0 Electron-Proton Scattering Experiment, *Physical Review Letters*, Vol. 95, No. 9, 092001 (5 pages).

Arnison, G. *et al.* (1983), Experimental Observation of Isolated Large Transverse Energy Electrons with Associated Missing Energy, *Physics Letters B*, Vol 122, No. 1, pp. 103-116.

Banner, M. *et al.* (1983), Observation of Single Isolated Electrons of High Transverse Momentum in Events with Missing Transverse Energy at the CERN pp Collider, *Physics Letters B*, Vol. 122, Nos. 5-6, pp. 476-485.

Bloom, E.D. *et al.* (1969), High-Energy Inelastic $e-p$ Scattering at $6^0$ and $10^0$, *Physical Review Letters*, Vol. 23, No. 16, pp. 930-934.

Breidenbach, M. *et al.* (1969), Observed Behavior of Highly Inelastic Electron-Proton Scattering, *Physical Review Letters*, Vol. 23, No. 16, pp. 935-939.

Butler, F. *et al.* (1993), Hadron Mass Predictions of the Valence Approximation to Lattice QCD, *Physical Review Letters*, Vol. 70, No. 19, pp. 2849-2852.

Cabibbo, N. (1963), Unitary Symmetry and Leptonic Decays, *Physical Review Letters*, Vol. 10, No. 12, pp. 531-533.

Casalbuoni, R. and Gatto, R. (1980), Subcomponent Models of Quarks and Leptons, *Physics Letters B*, Vol. 93, Nos. 1-2, pp. 47-52.

Chinowsky, W. and Steinberger, J. (1954), Absorption of Negative Pions in Deuterium: Parity of the Pion, *Physical Review*, Vol. 95, No. 6, pp. 1561-1564.

Christenson, J.H. *et al.* (1964), Evidence for the $2\pi$ Decay of the $K_2^0$ Meson, *Physical Review Letters*, Vol. 13, No. 4, pp. 138-140.

Cockcroft, J. and Walton, E. (1932), Experiments with High Velocity Positive Ions. II. The Disintegration of Elements by High Velocity Protons, *Proceedings of the Royal Society of London, Series A*, Vol. 137, No. 831, pp. 239-242.

Davies, C.T.H. *et al.* (2004), High-Precision Lattice QCD Confronts Experiment, *Physical Review Letters*, Vol. 92, No. 2, 022001 (5 pages).

Einstein, A. (1905), Ist die Trägheit eines Körpers von seinem Energieinhalt abhängig, *Annalen der Physik*, Vol. 18, No. 13, pp. 639-641.

Englert, F. and Brout, R. (1964), Broken Symmetry and the Mass of Gauge Vector Bosons, *Physical Review Letters*, Vol. 13, No. 9, pp. 321-323.

Evans, P.W. and Robson, B.A. (2006), Comparison of Quark Mixing in the Standard and Generation Models, *International Journal of Modern Physics E*, Vol. 15, No 3, pp. 617-625.

Franklin, A. (1986), *The Neglect of Experiment* (Cambridge University Press, Cambridge, U.K.).

Friedman, J.I. and Telegdi, V.L. (1957), Nuclear Emulsion Evidence for Parity Nonconservation in the Decay Chain $\pi^+ - \mu^+ - e^+$, *Physical Review*, Vol. 105, No. 5, pp. 1681-1682.

Garwin, R.L., Lederman, L.M. and Weinrich, M. (1957), Observations of the Failure of Conservation of Parity and Charge Conjugation in Meson Decays: the Magnetic Moment of the Free Muon, *Physical Review*, Vol. 105, No. 4, pp. 1415-1417.

Gell-Mann, M. and Ne'eman, Y. (1964), *The Eightfold Way*, (Benjamin, New York).

Gell-Mann, M. and Pais, A. (1955), Behavior of Neutral Particles under Charge Conjugation, *Physical Review*, Vol. 97, No. 5, pp. 1387-1389.

Glashow, S.L. (1961), Partial-Symmetries of Weak Interactions, *Nuclear Physics*, Vol. 22, pp. 579-588.

Gottfried, K. and Weisskopf, V.F. (1984), *Concepts of Particle Physics* Vol. 1 (Oxford University Press, New York).

Gross, D.J. and Wilczek, F. (1973), Ultraviolet Behavior of Non-Abelian Gauge Theories, *Physical Review Letters*, Vol. 30, No. 26, pp. 1343-1346.

Halzen, F. and Martin, A.D. (1984), *Quarks and Leptons: An Introductory Course in Modern Particle Physics* (John Wiley and Sons, New York).

Harari, H. (1979), A Schematic Model of Quarks and Leptons, *Physics Letters B*, Vol. 86, No. 1, pp. 83-86.

Harari, H. and Seiberg, N. (1981), A Dynamical Theory for the Rishon Model, *Physics Letters B*, Vol.98, No. 4, pp. 269-273.

Hasert *et al.* (1973), Observation of Neutrino-Like Interactions without Muon or Electron in the Gargamelle Neutrino Experiment, *Physics Letters B*, Vol. 46, No. 1, pp. 138-140.

Hasert *et al.* (1974), Observation of Neutrino-Like Interactions without Muon or Electron in the Gargamelle Neutrino Experiment, *Nuclear Physics B*, Vol. 73, No. 1, pp. 1-22.

Higgs, P.W. (1964), Broken Symmetries and the Masses of Gauge Bosons, *Physical Review Letters*, Vol. 13, No. 16, pp. 508-509.

Kabir, P.K. (1968), *The CP Puzzle: Strange Decays of the Neutral Kaon* (Academic Press, London).

Kleinknecht, K. (2003), *Uncovering CP Violation: Experimental Clarification in the Neutral K Meson and B Meson Systems* (Springer, Berlin).

Kobayashi, M. and Maskawa, T. (1973), CP-Violation in Renormalizable Theory of Weak Interaction, *Progress of Theoretical Physics*, Vol. 49, No. 2, pp. 652-657.

Landau, L.D. (1957), On the Conservation Laws for Weak Interactions, *Nuclear Physics*, Vol. 3, No.1, pp. 127-131.

Lee, T.D. and Yang, C.N. (1956), Question of Parity Conservation in Weak Interactions, *Physical Review*, Vol. 104, No. 1, pp. 254-258.

Lyre, H. (2008), Does the Higgs Mechanism Exist?, *International Studies in the Philosophy of Science*, Vol. 22, No. 2, pp. 119-133.

Mass, F.E. *et al.*, Evidence for Strange-Quark Contributions to the Nucleon's Form Factors at $Q^2 = 0.108\ (GeV/c)^2$, *Physical Review Letters*, Vol. 94, No. 15, 152001 (4 pages).

Morrison, A.D. and Robson, B.A. (2009), $2\pi$ Decay of the $K_L^0$ Meson without CP Violation, *International Journal of Modern Physics E*, Vol. 18, No. 9, pp. 1825-1830.

Politzer, H.D. (1973), Reliable Perturbative Results for Strong Interactions, *Physical Review Letters*, Vol. 30, No. 26, pp. 1346-1349.

Rainville, S. *et al.* (2005), World Year of Physics: A Direct Test of $E = mc^2$, *Nature*, Vol. 438, pp. 1096-1097.

Robson, B.A. (2002), A Generation Model of the Fundamental Particles, *International Journal of Modern Physics E*, Vol. 11, No. 6, pp. 555-566.

Robson, B.A. (2004), Relation between Strong and Weak Isospin, *International Journal of Modern Physics E*, Vol. 13, No. 5, pp. 999-1018.

Robson, B.A. (2005), A Generation Model of Composite Leptons and Quarks, *International Journal of Modern Physics E*, Vol. 14, No. 8, pp. 1151-1169.

Robson, B.A. (2008), The Generation Model and the Electroweak Connection, *International Journal of Modern Physics E*, Vol. 17, No. 6, pp. 1015-1030.

Robson, B.A. (2009), The Generation Model and the Origin of Mass, *International Journal of Modern Physics E*, Vol. 18, No. 8, pp. 1773-1780.

Robson, B.A. (2011a), A Quantum Theory of Gravity based on a Composite Model of Leptons and Quarks, *International Journal of Modern Physics E*, Vol. 20, No. 3, pp. 733-745.

Robson, B.A. (2011b), Parity of Charged Pions, *International Journal of Modern Physics E*, Vol. 20, No. 8, pp. 1677-1686.

Robson, B.A. (2011c), Parity of Neutral Pion, *International Journal of Modern Physics E*, Vol. 20, No. 9, pp. 1961-1965.

Salam, A. (1968) in *Elementary Particle Physics (Proceedings of the 8th Nobel Symposium)*, ed. Svartholm, N. (Almqvist and Wiksell, Stockholm), p. 367.

Shupe, M.A. (1979), A Composite Model of Leptons and Quarks, *Physics Letters B*, Vol. 86, No. 1, pp. 87-92.

Squires, E.J. (1980), QDD-a Model of Quarks and Leptons, *Physics Letters B*, Vol. 94, No. 1, pp. 54-56.

Squires, E.J. (1981), Some Comments on the Three-Fermion Composite Quark and Lepton Model, *Journal of Physics G*, Vol. 7, No. 4, pp. L47-L49.

t'Hooft, G. (1971a), Renormalization of Massless Yang-Mills Fields, *Nuclear Physics B*, Vol. 33, No. 1, pp. 173-199.

t'Hooft, G. (1971b), Renormalizable Lagrangians for Massive Yang-Mills Fields, *Nuclear Physics B*, Vol. 35, No. 1, pp. 167-188.

Veltman, M. (2003), *Facts and Mysteries in Elementary Particle Physics*, (World Scientific Publishing Company, Singapore).

Weinberg, S. (1967), A Model of Leptons, *Physical Review Letters*, Vol. 19, No. 21, pp. 1264-1266.

Wilczek, F. (2005) In Search of Symmetry Lost, *Nature*, Vol. 433, No. 3, pp. 239-247.

Wu, C.S. *et al.* (1957), Experimental Test of Parity Conservation in Beta Decay, *Physical Review*, Vol. 105, No. 4, pp. 1413-1415.

# 2

# Constraining the Couplings of a Charged Higgs to Heavy Quarks

A. S. Cornell
*National Institute for Theoretical Physics;*
*School of Physics, University of the Witwatersrand*
*South Africa*

## 1. Introduction

The Standard Model (SM) of particle physics has been an incredibly successful theory which has been confirmed experimentally many times, however, it still has some short-comings. As such physicists continue to search for models beyond the SM which might explain issues such as naturalness (the hierarchy problem). Among the possible discoveries that would signal the existence of these new physics models (among several) would be the discovery of a charged Higgs boson.

Recall that in the SM we have a single complex Higgs doublet, which through the Higgs mechanism, is responsible for breaking the Electroweak (EW) symmetry and endowing our particles with their mass. As a result we expect one neutral scalar particle (known as the Higgs boson) to emerge. Now whilst physicists have become comfortable with this idea, we have not yet detected this illusive Higgs boson. Furthermore, this approach leads to the hierarchy problem, where extreme fine-tuning is required to stabilise the Higgs mass against quadratic divergences. As such a simple extension to the SM, which is trivially consistent with all available data, is to consider the addition of extra $SU(2)$ singlets and/or doublets to the spectrum of the Higgs sector. One such extension shall be our focus here, that where we have two complex Higgs doublets, the so-called Two-Higgs Doublet Models (2HDMs). Such models, after EW symmetry breaking, will give rise to a charged Higgs boson in the physical spectrum. Note also that by having these two complex Higgs doublets we can significantly modify the Flavour Changing Neutral Current (FCNC) Higgs interactions in the large $\tan\beta$ region (where $\tan\beta \equiv v_2/v_1$, the ratio of the vacuum expectation values (vevs) of the two complex doublets).

Among the models which contain a second complex Higgs doublet one of the best motivated is the Minimal Supersymmetric Standard Model (MSSM). This model requires a second Higgs doublet (and its supersymmetric (SUSY) fermionic partners) in order to preserve the cancellation of gauge anomalies [1]. The Higgs sector of the MSSM contains two Higgs supermultiplets that are distinguished by the sign of their hypercharge, establishing an unambiguous theoretical basis for the Higgs sector. In this model the structure of the Higgs sector is constrained by supersymmetry, leading to numerous relations among Higgs masses and couplings. However, due to supersymmetry-breaking effects, all such relations are modified by loop-corrections, where the effects of supersymmetry-breaking can enter [1].

Thus, one can describe the Higgs-sector of the (broken) MSSM by an effective field theory consisting of the most general 2HDM, which is how we shall develop our theory in section 2.

Note that in a realistic model, the Higgs-fermion couplings must be chosen with some care in order to avoid FCNC [2, 3], where 2HDMs are classified by how they address this: In type-I models [4] there exists a basis choice in which only one of the Higgs fields couples to the SM fermions. In type-II [5, 6], there exists a basis choice in which one Higgs field couples to the up-type quarks, and the other Higgs field couples to the down-type quarks and charged leptons. Type-III models [7] allow both Higgs fields to couple to all SM fermions, where such models are viable only if the resulting FCNC couplings are small.

Once armed with a model for a charged Higgs boson, we must determine how this particle will manifest and effect our experiments. Of the numerous channels, both direct and indirect, in which its presence could have a profound effect, one of the most constraining are those where the charged Higgs mediates tree-level flavour-changing processes, such as $B \to \tau\nu$ and $B \to D\tau\nu$ [8]. As these processes have already been measured at $B$-factories, they will provide us with very useful indirect probes into the charged Higgs boson properties. Furthermore, with the commencement of the Large Hadron Collider (LHC) studies involving the LHC environment promise the best avenue for directly discovering a charged Higgs boson. As such we shall determine the properties of the charged Higgs boson using the following processes:

- **LHC:** $pp \to t(b)H^+$: through the decays $H^\pm \to \tau\nu$, $H^\pm \to tb$ ($b - t - H^\pm$ coupling).
- **$B$-factories:** $B \to \tau\nu$ ($b - u - H^\pm$ coupling), $B \to D\tau\nu$ ($b - c - H^\pm$ coupling).

The processes mentioned above have several common characteristics with regard to the charged Higgs boson couplings to the fermions. Firstly, the parameter region of $\tan\beta$ and the charged Higgs boson mass covered by charged Higgs boson production at the LHC ($pp \to t(b)H^+$) overlaps with those explored at $B$-factories. Secondly, these processes provide four independent measurements to determine the charged Higgs boson properties. With these four independent measurements one can in principle determine the four parameters related to the charged Higgs boson couplings to $b$-quarks, namely $\tan\beta$ and the three generic couplings related to the $b - i - H^\pm$ ($i = u, c, t$) vertices. In our analysis we focus on the large $\tan\beta$-region [9], where one can neglect terms proportional to $\cot\beta$, where at tree-level the couplings to fermions will depend only on $\tan\beta$ and the mass of the down-type fermion involved. Hence, at tree-level, the $b - i - H^\pm$ ($i = u, c, t$) vertex is the same for all the three up-type generations. This property is broken by loop corrections to the charged Higgs boson vertex.

Our strategy in this pedagogical study will be to determine the charged Higgs boson properties first through the LHC processes. Note that the latter have been extensively studied in many earlier works (see Ref.[10], for example) with the motivation of discovering the charged Higgs boson in the region of large $\tan\beta$. We shall assume that the charged Higgs boson is already observed with a certain mass. Using the two LHC processes as indicated above, one can then determine $\tan\beta$ and the $b - t - H^\pm$ coupling. Having an estimate of $\tan\beta$ one can then study the $B$-decays and try to determine the $b - (u/c) - H^\pm$ couplings from $B$-factory measurements. This procedure will enable us to measure the charged Higgs boson couplings to the bottom quark and up-type quarks [11].

The chapter will therefore be organised in the following way: In Section 2 we shall discuss the model we have considered for our analysis. As we shall use an effective field theory derived from the MSSM, we will also introduce the relevant SUSY-QCD and higgsino-stop loop correction factors to the relevant charged Higgs boson fermion couplings. Using this formalism we shall study in section 3 the possibility of determining the charged Higgs boson properties at the LHC using $H^{\pm} \to \tau\nu$ and $H^{\pm} \to tb$. In Section 4 we shall present the results of $B$-decays, namely $B \to \tau\nu$ and $B \to D\tau\nu$, as studied in Ref.[8]. Finally, we shall combine the $B$-decay results with our LHC simulations to determine the charged Higgs boson properties (such as its mass, $\tan\beta$ and SUSY loop correction factors) and give our conclusions.

## 2. Effective Lagrangian for a charged Higgs boson

In this section we shall develop the general form of the effective Lagrangian for the charged Higgs interactions with fermions. As already discussed in the introduction of this chapter, at tree-level the Higgs sector of the MSSM is of the same form as the type-II 2HDM, also in (at least in certain limits of) those of type-III. In these 2HDMs the consequence of this extended Higgs sector is the presence of additional Higgs bosons in the physics spectrum. In the MSSM we will have 5 Higgs bosons, three neutral and two charged.

### 2.1 The MSSM charged Higgs

We shall begin by recalling that we require at least two Higgs doublets in SUSY theories, where in the SM the Higgs doublet gave mass to the leptons and down-type quarks, whilst the up-type quarks got their mass by using the charge conjugate (as was required to preserve all gauge symmetries in the Yukawa terms). In the SUSY case the charge conjugate cannot be used in the superpotential as it is part of a supermultiplet. As such the simplest solution is to introduce a second doublet with opposite hypercharge. So our theory will contain two chiral multiplets made up of our two doublets $H_1$ and $H_2$ and corresponding higgsinos $\widetilde{H}_1$ and $\widetilde{H}_2$ (fields with a tilde ($\widetilde{\ }$) denote squarks and sleptons); in which case the superpotential in the MSSM is:

$$W = -H_1 D^c \mathbf{y}_d Q + H_2 U^c \mathbf{y}_u Q - H_1 E^c \mathbf{y}_e L + \mu H_1 H_2 \,. \tag{1}$$

The components of the weak doublet fields are denoted as:

$$H_1 = \begin{pmatrix} H_1^0 \\ H_1^- \end{pmatrix} , \; H_2 = \begin{pmatrix} H_2^+ \\ H_2^0 \end{pmatrix} , \; Q = \begin{pmatrix} U \\ D \end{pmatrix} , \; L = \begin{pmatrix} N \\ E \end{pmatrix} . \tag{2}$$

The quantum numbers of the $SU(3) \times SU(2) \times U(1)$ gauge groups for $H_1$, $H_2$, $Q$, $L$, $D^c$, $U^c$, $E^c$ are $(\mathbf{1},\mathbf{2},-1)$, $(\mathbf{1},\mathbf{2},1)$, $(\mathbf{3},\mathbf{2},\frac{1}{3})$, $(\mathbf{1},\mathbf{2},-1)$, $(\mathbf{3},\mathbf{1},\frac{2}{3})$, $(\mathbf{3},\mathbf{1},-\frac{4}{3})$, $(\mathbf{1},\mathbf{1},2)$; where the gauge and family indices were eliminated in Eq.(1). For example $\mu H_1 H_2 = \mu (H_1)_\alpha (H_2)_\beta \epsilon^{\alpha\beta}$ with $\alpha, \beta = 1,2$ being the $SU(2)_L$ isospin indices and $H_1 D^c \mathbf{y}_d Q = (H_1)_\beta D^{ci}_a (\mathbf{y}_d)_i^j Q^a_{j\alpha} \epsilon^{\alpha\beta}$ with $i,j = 1,2,3$ as the family indices and $a = 1,2,3$ as the colour indices of $SU(3)_c$. As in the SM the Yukawas $\mathbf{y}_d$, $\mathbf{y}_u$ and $\mathbf{y}_e$ are $3 \times 3$ unitary matrices.

Note that Eq.(1) does not contain terms with $H_1^*$ or $H_2^*$, consistent with the fact that the superpotential is a holomorphic function of the supermultiplets. Yukawa terms like $\bar{U} Q H_1^*$, which are usually present in non-SUSY models, are excluded by the invariance under the supersymmetry transformation.

The soft SUSY breaking masses and trilinear SUSY breaking terms ($A$-term) are given by:

$$\begin{aligned}\mathcal{L}_{\rm soft} = & -\widetilde{Q}_{\rm L}^{\dagger} M_{\widetilde{Q}_{\rm L}}^2 \widetilde{Q}_{\rm L} - \widetilde{U}_{\rm R}^{\dagger} M_{\widetilde{U}_{\rm R}}^2 \widetilde{U}_{\rm R} - \widetilde{D}_{\rm R}^{\dagger} M_{\widetilde{D}_{\rm R}}^2 \widetilde{D}_{\rm R} - \widetilde{L}_{\rm L}^{\dagger} M_{\widetilde{L}_{\rm L}}^2 \widetilde{L}_{\rm L} - \widetilde{E}_{\rm R}^{\dagger} M_{\widetilde{E}_{\rm R}}^2 \widetilde{E}_{\rm R} \\ & + H_1 \widetilde{D}_{\rm R}^{\dagger} \mathbf{A}_d \widetilde{Q}_{\rm L} - H_2 \widetilde{U}_{\rm R}^{\dagger} \mathbf{A}_u \widetilde{Q}_{\rm L} + H_1 \widetilde{E}_{\rm R}^{\dagger} \mathbf{A}_e \widetilde{L}_{\rm L} + \text{h.c.} \end{aligned} \quad (3)$$

Let us first discuss the simplest case where soft breaking masses are proportional to a unit matrix in the flavour space, and $\mathbf{A}_u$, $\mathbf{A}_d$ and $\mathbf{A}_e$ are proportional to Yukawa couplings. Their explicit forms being:

$$\begin{aligned} & M^2_{\widetilde{Q}_{{\rm L}ij}} = a_1 \widetilde{M}^2 \delta_{ij}\,,\; M^2_{\widetilde{U}_{{\rm R}ij}} = a_2 \widetilde{M}^2 \delta_{ij}\,,\; M^2_{\widetilde{D}_{{\rm R}ij}} = a_3 \widetilde{M}^2 \delta_{ij}\,,\; M^2_{\widetilde{L}_{{\rm L}ij}} = a_4 \widetilde{M}^2 \delta_{ij}\,, \\ & M^2_{\widetilde{E}_{{\rm R}ij}} = a_5 \widetilde{M}^2 \delta_{ij}\,,\; \mathbf{A}_{uij} = A_u \mathbf{y}_{uij}\,,\; \mathbf{A}_{dij} = A_d \mathbf{y}_{dij}\,,\; \mathbf{A}_{eij} = A_e \mathbf{y}_{eij}\,, \end{aligned} \quad (4)$$

where $a_i (i = 1 - 5)$ are real parameters.

At tree-level the Yukawa couplings have the same structure as the above superpotential, namely, $H_1$ couples to $D^c$ and $E^c$, and $H_2$ to $U^c$. On the other hand, different types of couplings are induced when we take into account SUSY breaking effects through one-loop diagrams. The Lagrangian of the Yukawa sector can be written as:

$$\begin{aligned}\mathcal{L}_{\rm Yukawa} = & -H_1 \overline{D}_{\rm R} \mathbf{y}_d Q_{\rm L} + H_2 \overline{U}_{\rm R} \mathbf{y}_u Q_{\rm L} - H_1 \overline{E}_{\rm R} \mathbf{y}_e L_{\rm L} - i\sigma_2 H_2^* \overline{D}_{\rm R} \Delta \mathbf{y}_d Q_{\rm L} \\ & + i\sigma_2 H_1^* \overline{U}_{\rm R} \Delta \mathbf{y}_u Q_{\rm L} - i\sigma_2 H_2^* \overline{E}_{\rm R} \Delta \mathbf{y}_e L_{\rm L} + \text{h.c.}\,, \end{aligned} \quad (5)$$

where $\Delta \mathbf{y}_d$, $\Delta \mathbf{y}_u$, and $\Delta \mathbf{y}_e$ are one-loop induced coupling constants, and we recall that gauge indices have been suppressed; for example $\sigma_2 H_2^* \overline{D}_{\rm R} \Delta \mathbf{y}_d Q_{\rm L} = (\sigma_2)^{\alpha\beta} (H_2^*)_\beta (\overline{D}_{\rm R})_a^i (\Delta \mathbf{y}_d)_i^j (Q_{\rm L})_{j\alpha}^a$. From the above Yukawa couplings, we can derive the quark and lepton mass matrices and their charged Higgs couplings. For the quark sector, we get

$$\begin{aligned}\mathcal{L}_{\rm quark} = & -\frac{v}{\sqrt{2}} \cos\beta \overline{D}_{\rm R} \mathbf{y}_d [1 + \tan\beta \Delta_{m_d}] D_{\rm L} + \sin\beta H^- \overline{D}_{\rm R} \mathbf{y}_d [1 - \cot\beta \Delta_{m_d}] U_{\rm L} \\ & -\frac{v}{\sqrt{2}} \sin\beta \overline{U}_{\rm R} \mathbf{y}_u [1 - \cot\beta \Delta_{m_u}] U_{\rm L} + \cos\beta H^+ \overline{U}_{\rm R} \mathbf{y}_u [1 + \tan\beta \Delta_{m_u}] D_{\rm L} + \text{h.c.}\,, \end{aligned} \quad (6)$$

where we define $\Delta_{m_d} (\Delta_{m_u})$ as $\Delta_{m_d} \equiv \mathbf{y}_d^{-1} \Delta \mathbf{y}_d$ ($\Delta_{m_u} \equiv \mathbf{y}_u^{-1} \Delta \mathbf{y}_u$), and $v \simeq 246$GeV. Notice that $\Delta \mathbf{y}_d$ is proportional to $\mathbf{y}_d$ or $\mathbf{y}_d \mathbf{y}_u^\dagger \mathbf{y}_u$ in this case. We then rotate the quark bases as follows:

$$U_{\rm L} = V_{\rm L}(Q) U'_{\rm L}\,,\; D_{\rm L} = V_{\rm L}(Q) V_{\rm CKM} D'_{\rm L}\,,\; U_{\rm R} = V_{\rm R}(U) U'_{\rm R}\,,\; D_{\rm R} = V_{\rm R}(D) D'_{\rm R}\,, \quad (7)$$

where the fields with a prime ($'$) are mass eigenstates. In this basis, the down-type quark Lagrangian is given by

$$\begin{aligned}\mathcal{L}_{\rm D-quark} = & -\frac{v}{\sqrt{2}} \cos\beta \overline{D}_{\rm R}{}' V_{\rm R}^\dagger(D) \mathbf{y}_d V_{\rm L}(Q) \widehat{R}_d V_{\rm CKM} D'_{\rm L} \\ & + \sin\beta H^- \overline{D}_{\rm R}{}' V_{\rm R}^\dagger(D) \mathbf{y}_d V_{\rm L}(Q) U'_{\rm L} + \text{h.c.}\,, \end{aligned} \quad (8)$$

where $\widehat{R}_d \equiv 1 + \tan\beta \widehat{\Delta}_{m_d}$ and $\widehat{\Delta}_{m_d} \equiv V_{\rm L}^\dagger(Q) \Delta_{m_d} V_{\rm L}(Q)$. Hereafter, a matrix with a hat ($\widehat{\ }$) represents a diagonal matrix. Since the down-type diagonal mass term is given by

$$\widehat{M}_d \equiv \frac{v}{\sqrt{2}} \cos\beta V_{\rm R}^\dagger(D) \mathbf{y}_d V_{\rm L}(Q) \widehat{R}_d V_{\rm CKM}\,, \quad (9)$$

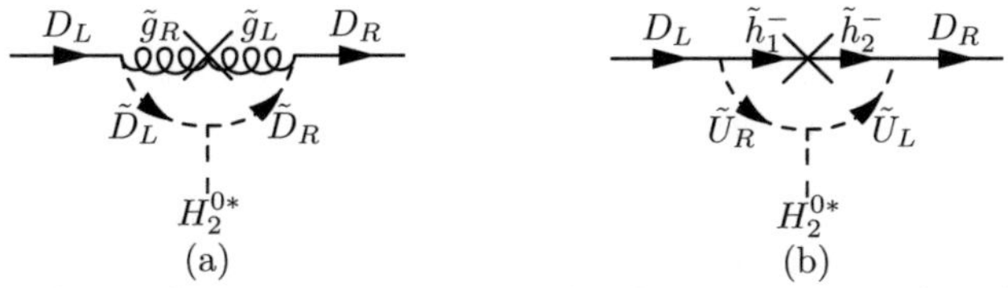

Fig. 1. Non-holomorphic radiative corrections to the down-type quark Yukawa couplings induced by (a) gluino $\widetilde{g}_{\rm L,R}$ and (b) charged higgsino $\widetilde{h}^-_{1,2}$.

we obtain the following Lagrangian for down-type quarks.

$$\mathcal{L}_{\rm D-quark} = -\overline{D}_{\rm R}{}' \widehat{M}_d D_{\rm L}' + \frac{\sqrt{2}}{v} \tan\beta H^- \overline{D}_{\rm R}{}' \widehat{M}_d V_{\rm CKM}^\dagger \widehat{R}_d^{-1} U_{\rm L}' + {\rm h.c.} \tag{10}$$

The corresponding corrections to the up-type couplings can be calculated from Eq.(6). Since we are interested in the large $\tan\beta$ case, these corrections are very small. In the following we neglect such corrections, and the Lagrangian for the up-type quarks is given as follows:

$$\mathcal{L}_{\rm U-quark} = -\overline{U}_{\rm R}{}' \widehat{M}_u U_{\rm L}' + \frac{\sqrt{2}}{v} \cot\beta H^+ \overline{U}_{\rm R}{}' \widehat{M}_u V_{\rm CKM} D_{\rm L}' + {\rm h.c.} \tag{11}$$

For the case of the charged-lepton, we can derive the relevant parts of the Lagrangian in a similar way to the case of the down-type quark by choosing an appropriate basis choice.

In the present case with Eqs.(4) $\widehat{\Delta}_{m_d}$ receives contributions from gluino and down-type squark, and higgsino and up-type squark diagrams. The explicit form is given as follows:

$$\widehat{\Delta}_{m_d} = \widehat{\mathbf{E}}_{\widetilde{g}} + \widehat{\mathbf{E}}_{\widetilde{h}} \, , \tag{12}$$

where

$$\widehat{\mathbf{E}}_{\widetilde{g}} \equiv \frac{2\alpha_s}{3\pi} \mathbf{1} \mu^* M_{\widetilde{g}} I[M_{\widetilde{g}}, M_{\widetilde{D}_{\rm L}}, M_{\widetilde{D}_{\rm R}}] \, , \tag{13}$$

$$\widehat{\mathbf{E}}_{\widetilde{h}} \equiv -\frac{\mu}{16\pi^2} A_{\rm u} |\widehat{\mathbf{y}}_{\rm u}|^2 I[M_{\widetilde{h}}, M_{\widetilde{U}_{\rm L}}, M_{\widetilde{U}_{\rm R}}] \, , \tag{14}$$

$$I[a,b,c] = \frac{a^2 b^2 \ln\frac{a^2}{b^2} + b^2 c^2 \ln\frac{b^2}{c^2} + c^2 a^2 \ln\frac{c^2}{a^2}}{(a^2-b^2)(b^2-c^2)(a^2-c^2)} \, . \tag{15}$$

$\widehat{\mathbf{E}}_{\widetilde{g}}$ and $\widehat{\mathbf{E}}_{\widetilde{h}}$ are gluino and charged higgsino contributions shown in Fig.1(a) and (b) respectively. Note that these corrections for Yukawa couplings are calculated in the unbroken phase of $SU(2) \times U(1)$.

Up to now we have assumed all squark mass matrices are proportional to a unit matrix at the EW scale, as shown in Eqs.(4). However, models with Minimal Flavour Violation (MFV) correspond to more general cases. For instance, the assumption of Eqs.(4) is not satisfied in minimal supergravity, where all squarks have a universal mass at the Planck scale, not at the EW scale. In Ref.[8] they derive the charged Higgs coupling in a more general case of MFV. Namely the squark mass matrix is taken to be

$$\begin{aligned} M^2_{\widetilde{Q}_{\rm L}} &= [a_1 \mathbf{1} + b_1 \mathbf{y}_u^\dagger \mathbf{y}_u + b_2 \mathbf{y}_d^\dagger \mathbf{y}_d] \widetilde{M}^2 \, , \\ M^2_{\widetilde{U}_{\rm R}} &= [a_2 \mathbf{1} + b_5 \mathbf{y}_u \mathbf{y}_u^\dagger] \widetilde{M}^2 \, , \\ M^2_{\widetilde{D}_{\rm R}} &= [a_3 \mathbf{1} + b_6 \mathbf{y}_d \mathbf{y}_d^\dagger] \widetilde{M}^2 \, . \end{aligned} \tag{16}$$

The final results of the charged Higgs coupling being given by

$$\mathcal{L}_{H^\pm} \approx \frac{\sqrt{2}}{v}\tan\beta H^- \overline{D}_{Ri}' \frac{\widehat{M}_{di}}{1+[E_{\tilde{g}}^{(i)}]\tan\beta} V^\dagger_{\mathrm{CKM}ij} U'_{Lj} + \text{h.c.} \quad \text{for } (i,j) = (1,1),(1,2),(2,1),(2,2), \tag{17}$$

$$\mathcal{L}_{H^\pm} \approx \frac{\sqrt{2}}{v}\tan\beta H^- \overline{D}_{Ri}' \frac{\widehat{M}_{di}}{1+[E_{\tilde{g}}^{(i)} - E'^{(ij)}_{\tilde{g}}]\tan\beta} V^\dagger_{\mathrm{CKM}ij} U'_{Lj} + \text{h.c.} \quad \text{for } (i,j) = (3,1),(3,2), \tag{18}$$

$$\mathcal{L}_{H^\pm} \approx \frac{\sqrt{2}}{v}\tan\beta H^- \overline{D}_{Ri}' \frac{\widehat{M}_{di}}{1+E_{\tilde{g}}^{(i)}\tan\beta} \frac{1+[E_{\tilde{g}}^{(3)}+E_{\tilde{h}}^{(33)}]\tan\beta}{1+[E_{\tilde{g}}^{(i)}+E_{\tilde{h}}^{(33)}+E'^{(ij)}_{\tilde{g}}+E_{\tilde{h}}^{(i3)}+E'^{(i33)}_{\tilde{h}}]\tan\beta} \times V^\dagger_{\mathrm{CKM}ij} U'_{Lj} + \text{h.c.} \quad \text{for } (i,j) = (1,3),(2,3), \tag{19}$$

$$\mathcal{L}_{H^\pm} \approx \frac{\sqrt{2}}{v}\tan\beta H^- \overline{D}_{Ri}' \frac{\widehat{M}_{di}}{1+[E_{\tilde{g}}^{(i)}+E_{\tilde{h}}^{(i3)}]\tan\beta} V^\dagger_{\mathrm{CKM}ij} U'_{Lj} + \text{h.c.} \quad \text{for } (i,j) = (3,3). \tag{20}$$

The functions $E_{\tilde{g}}^{(i)}$, etc. are listed in Ref.[8]. In deriving these results only the $y_t$ in the up-type Yukawa coupling in loop diagrams was kept and use made of the hierarchy of the CKM matrix elements. See Ref.[8] for details. Notice that the above results do not depend on the relationship between the $A$-terms and the Yukawa couplings, since only the $y_t$ in loop diagrams was kept, even though Eqs.(4) are assumed.

### 2.2 Couplings to the bottom quark

From Eq.(10) and now under the assumption of MFV, we know that trilinear couplings are in general proportional to the original Yukawa couplings. We shall therefore label the components of the diagonal matrix $\widehat{R}_d^{-1} = \text{diag}\left[R_{11}^{-1}, R_{22}^{-1} R_{33}^{-1}\right]$, where the three diagonal values of $\widehat{R}_d^{-1}$ represent the couplings of a charged Higgs boson to the bottom quark and the three up-type quarks. At tree-level, these three couplings are equal, $R_{11}^{-1} = R_{22}^{-1} = R_{33}^{-1} = 1$, where this equality is broken to some extent by loop corrections to the charged Higgs vertex, and $\widehat{R}_d$ can then be written as:

$$\widehat{R}_d = 1 + \tan\beta\hat{\Delta}_{m_d} \,. \tag{21}$$

In the forth-coming analysis we have kept the $\mathcal{O}(\alpha_s)$ SUSY-QCD corrections and SUSY loop corrections associated with the Higgs-top Yukawa couplings (as discussed in the previous subsection) and have neglected the subleading EW corrections of the order $\mathcal{O}(g^2)$ as given in Ref.[12].[1] Therefore, they then depend upon the higgsino-mass parameter $\mu$, the up-type trilinear couplings $A$, and the bino, bottom and top squark masses. As argued in Ref.[8] the higgsino-diagram contributions can be neglected in $R_{11}^{-1}$ and $R_{22}^{-1}$, so that to a very good approximation $R_{11}^{-1} \approx R_{22}^{-1}$. As an illustration, we show in Fig.2 the dependence of the SUSY corrections on $\tan\beta$ for some illustrative SUSY parameters. These corrections can alter the tree-level values significantly, although low-energy data (e.g. from $b \to s\gamma$, $B - \bar{B}$ mixing,

[1] For an alternative definition, in which SUSY loop effects are assigned to the CKM matrix, see Ref.[13]

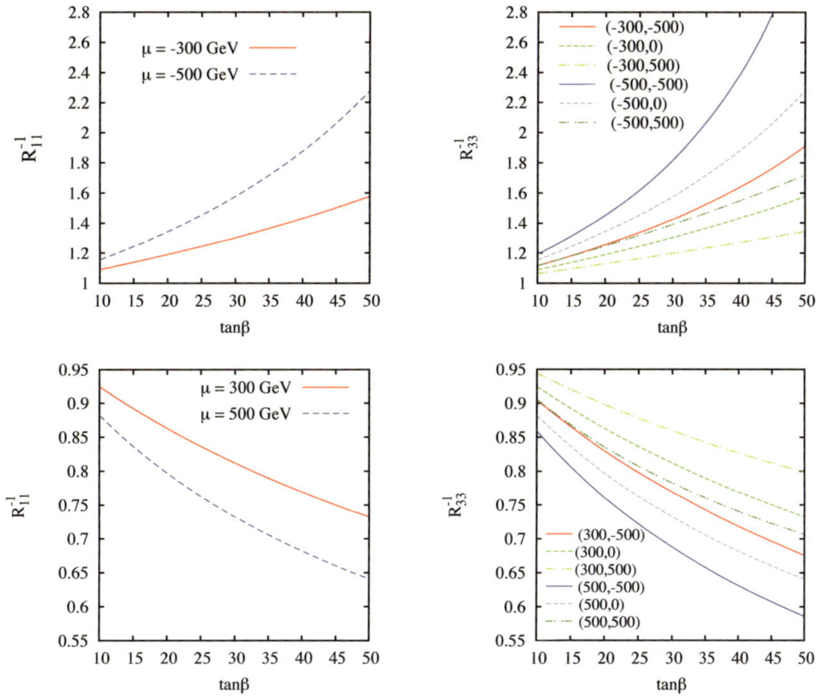


Fig. 2. Dependence of the general couplings $R_{ii}^{-1}$ on $\tan\beta$ in the exemplary case of the MSSM for various values of the higgsino mass parameter $\mu$ and the up-type trilinear coupling $A$. The left-hand plots are for $R_{11}^{-1} = R_{22}^{-1}$, while those on the right are for $R_{33}^{-1}$. We present the case of negative $\mu$ in the top panels and for positive $\mu$ below. The other SUSY parameters are $M_{\tilde{g}}$ = 800 GeV and $M_{\tilde{b}_1} = M_{\tilde{t}_1}$ = 500 GeV. We have also assumed $M_{\tilde{t}_L} = M_{\tilde{t}_R}$ and $M_{\tilde{b}_L} = M_{\tilde{b}_R}$. The legends in the right top and right bottom panels correspond to $(\mu, A)$ in GeV.

$B \to \mu\mu$ and $b \to s\mu\mu$) restricts the admissible parameter space [14]. In addition, it can be observed that the higgsino corrections are proportional to the up-type Yukawa couplings and hence can be substantial for diagrams involving the top quark as an external fermion line. This effectively implies that $R_{33}^{-1}$ can differ substantially from $R_{11}^{-1}$, where for certain SUSY scenarios, as shown in Fig.2, we observe that $R_{33}^{-1}$ can differ from $R_{11}^{-1}$ by more than 30%. This difference could be observed at the LHC for processes that depend on $R_{33}^{-1}$ when compared with the results of $B$-factories for processes that depend on $R_{11}^{-1}$. We remind the reader that the effective couplings are invariant under a rescaling of all SUSY masses and may indeed be the first observable SUSY effect, as long as the heavy Higgs bosons are light enough. The situation is similar in other models predicting a charged Higgs boson, such as those with a Peccei-Quinn symmetry, spontaneous $CP$ violation, dynamical symmetry breaking, or those based on $E_6$ superstring theories, but these have usually been studied much less with respect to the constraints imposed by low-energy data. In the remainder of this work, we shall thus treat the diagonal entries of $\widehat{R}_d^{-1}$ as model-independent free parameters in our simulations and numerics, but we will assume that $R_{11}^{-1} \approx R_{22}^{-1}$. Note that the corresponding corrections to the up-type couplings are suppressed by $\cot\beta$ and hence can be neglected in our analysis.

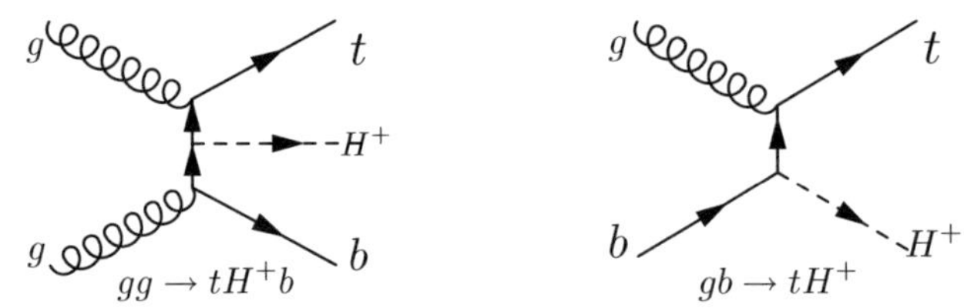


Fig. 3. The charged Higgs production at the LHC through the $gg \to tbH^{\pm}$ process, the $gb \to tH^{\pm}$ process, and there will also be parton level processes. The inclusive cross-section is the sum of these contributions, after the subtraction of common terms.

## 3. The $H^{\pm}$ decay channels at the LHC

With the theory for a charged Higgs coupling to heavy quarks now developed, we shall now consider the case where the charged Higgs boson is heavier than the top quark mass. Our reasoning for doing this, in this illustrative example, is that experimental searches have already placed a lower limit on the mass of a charged Higgs, including LEP, which set a limit of $m_{H^{\pm}} > 78.6$ GeV [15]. Note that within the MSSM, the charged Higgs mass is constrained by the pseudo-scalar Higgs mass and $W$-boson mass at tree level, with only moderate higher-order corrections, resulting in $m_{H^{\pm}} \gtrsim 120$ GeV. Furthermore, the Tevatron constrains (in several different MSSM scenarios) $m_{H^{\pm}} \gtrsim 150$ GeV [16], and at the LHC ATLAS has so far found (for $\tan\beta > 22$) $m_{H^{\pm}} > 140$ GeV [17] and CMS $m_{H^{\pm}} \gtrsim 160$ GeV [18].

As such, with $m_{H^{\pm}} \gtrsim m_t$, the production mechanism at the LHC shall be the associated production $pp \to tbH^{\pm} + X$ (the main production mechanisms are then $gg \to tbH^{\pm}$, $gb \to tH^{\pm}$ and the parton level processes, as shown in Fig.3[19]), with alternative production mechanisms like quark-antiquark annihilation, $q\bar{q} \to H^+H^-$[20] and $H^{\pm}+$ jet production, associated production with a $W$ boson, $q\bar{q} \to H^{\pm}W^{\mp}$[21], or Higgs pair production having suppressed rates. Note that some of the above production processes may be enhanced in models with non-MFV, which we shall not consider here.

Once produced, it is expected that the decay channel $H^+ \to \tau\nu$ shall be the primary discovery channel for the charged Higgs boson. Recall that we shall consider the large $\tan\beta$ region, where the branching ratios of charged decays into SM particles is given in Fig.4[10]. For $\tan\beta = 40$ the branching ratio for $H^+ \to tb$ is also quite high, we shall therefore consider both decay channels here. Note that we have assumed a heavy SUSY spectrum, such that the charged Higgs will decay only into SM particles for the maximal stop mixing scenario. For low values of $\tan\beta$, below the top quark mass, the main decay channels are $H^{\pm} \to \tau^{\pm}\nu_{\tau}$, $c\bar{s}$, $Wh^0$ and $t^*b$.

As such we shall now simulate the charged Higgs boson in the LHC environment with as much care as is possible, where we have included QCD corrections, as well as fully analysing the $H^+ \to tb$ mode. We should note though that of the main production mechanisms in Fig.3, there will be a partial overlap when the $gb \to tH^{\pm}$ is obtained from the $gg \to tbH^{\pm}$ by a gluon splitting into a $b$-quark pair. The summing of both contributions must be done with care, so as to avoid double counting, as we shall now discuss in greater detail.

### 3.1 The resolution of double-counting and the normalisation of the cross-section

From the associated production $pp \to tbH^{\pm} + X$, two different mechanisms can be employed to calculate the production cross-section. The first is the four flavour scheme with no $b$ quarks

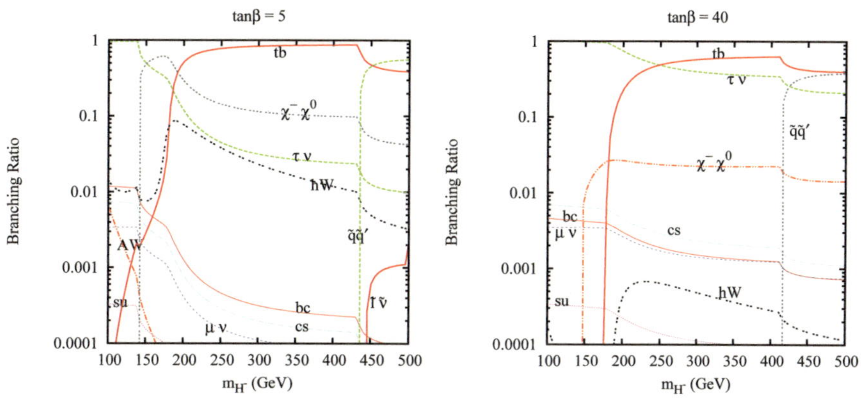


Fig. 4. The branching ratios of charged decays into SM particles as a function of $m_{H^\pm}$, for $\tan\beta = 5$ (left panel), and $\tan\beta = 40$ (right panel)[10].

in the initial state, the lowest order QCD production processes are gluon-gluon fusion and quark-antiquark annihilation, $gg \to tbH^\pm$ and $q\bar{q} \to tbH^\pm$ respectively. Note that potentially large logarithms $\propto \ln(\mu_F/m_b)$, arising from the splitting of incoming gluons into nearly collinear $b\bar{b}$ pairs, can be summed to all orders in perturbation theory by introducing bottom parton densities. This then defines the five flavour scheme. The use of bottom distribution functions is based on the approximation that the outgoing $b$ quark is at small transverse momentum and massless, and the virtual $b$ quark is quasi on-shell. In this scheme, the leading order process for the inclusive $tbH^\pm$ cross-section is gluon-bottom fusion, $gb \to tH^\pm$. The corrections to $gb \to tH^\pm$ and tree-level processes $gg \to tbH^\pm$ and $q\bar{q} \to tbH^\pm$. To all orders in perturbation theory the four and five flavour schemes are identical, but the way of ordering the perturbative expansion is different, and the results do now match exactly at finite order.

As such, in order to resolve the double-counting problem during event generation we use MATCHIG[22] as an external process to PYTHIA6.4.11[23]. In this program, when the $gb \to tH^-$ ($g\bar{b} \to \bar{t}H^+$) process is generated, there will be an accompanying outgoing $\bar{b}$ ($b$) quark. For low transverse momenta of this accompanying $b$ quark, this process, including initial state parton showers, describes the cross-section well. However, for large transverse momentum of the accompanying $b$-quark one instead uses the exact matrix element of the $gg \to t\bar{b}H^-$ ($gg \to \bar{t}bH^+$) process. Whilst for low transverse momenta, this process can be described in terms of the gluon splitting to $b\bar{b}$ times the matrix element of the $gb \to tH^\pm$ process. As was shown in Ref.[24], for low transverse momenta ($\lesssim$ 100GeV) the $gg \to t\bar{b}H^\pm$ approach underestimates the differential cross-section. Therefore, when the accompanying $b$-quark is observed, it is necessary to use both the $g\bar{b} \to tH^\pm$ and the $gg \to t\bar{b}H^\pm$ processes together, appropriately matched to remove the double-counting.

To do this MATCHIG defines a double-counting term $\sigma_{DC}$, given by the part of the $gg \to t\bar{b}H^\pm$ process which is already included in the $g\bar{b} \to tH^\pm$ process. This term is then subtracted from the sum of the cross-sections of the two processes. The double-counting term is given by the leading contribution of the $b$ quark density as:

$$\sigma_{DC} = \int dx_1 dx_2 \left[ g(x_1, \mu_F) b'(x_2, \mu_F) \frac{d\hat{\sigma}_{2\to 2}}{dx_1 dx_2}(x_1, x_2) + x_1 \leftrightarrow x_2 \right] , \tag{22}$$

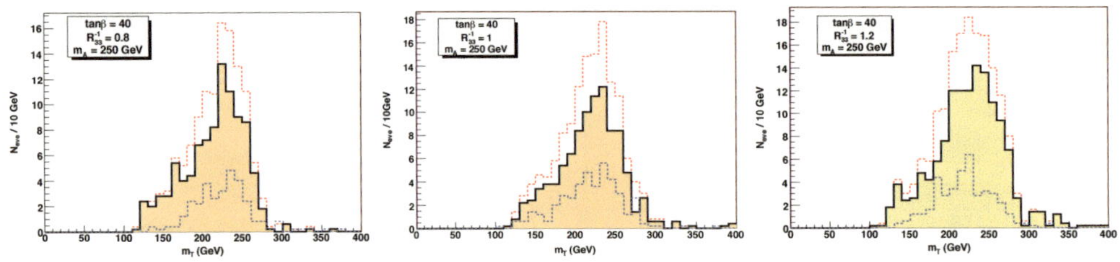


Fig. 5. Plots of the transverse mass of the charged Higgs in $H \to \tau\nu$ for a luminosity of $300\text{fb}^{-1}$ scaled to $30\text{fb}^{-1}$. The three lines in each plot correspond to positive events (the dotted red lines), negative events (dotted and blue) and matched events (shaded portion and black). The three graphs corresponds to three different values of $R^{-1}$ as indicated in each plot.

where $b'(x, \mu_F^2)$ is the leading order $b$-quark density given by [22]:

$$b'(x, \mu_F^2) \approx \frac{\alpha_s}{2\pi} \log \frac{\mu_F^2}{m_b^2} \int \frac{dz}{z} P_{qg}(z)\, g\left(\frac{x}{z}, \mu_F^2\right) , \tag{23}$$

with $P_{qg}$ the $g \to q\bar{q}$ splitting function, $g(x, \mu_F^2)$ the gluon density function, $\mu_F$ the factorization scale and $z$ the longitudinal gluon momentum fraction taken by the $b$-quark.

Including kinematic constraints due to finite center of mass energy (CM) and finite $b$ quark mass, the resulting expression for the double-counting term can be written as [24]:

$$\sigma_{\text{DC}} = \int_{\tau_{\min}}^{1} \frac{d\tau}{\tau} \int_{\frac{1}{2}\log\tau}^{-\frac{1}{2}\log\tau} dy^* \frac{\pi}{\hat{s}} \int_{-1}^{1} \frac{\beta_{34}}{2} d(\cos\hat{\theta})\, |\mathcal{M}_{2\to2}|^2 \, \frac{\alpha_s(\mu_R^2)}{2\pi}$$
$$\times \left[ \int_{x_1}^{z_{\max}} dz P_{qg}(z) \int_{Q^2_{\min}}^{Q^2_{\max}} \frac{d(Q^2)}{Q^2 + m_b^2} \frac{x_1}{z} g\left(\frac{x_1}{z}, \mu_F^2\right) x_2 g(x_2, \mu_F^2) + x_1 \leftrightarrow x_2 \right] . \tag{24}$$

Here $\mathcal{M}_{2\to2}$ is the matrix element for the $g\bar{b} \to tH^{\pm}$ process, $\mu_F$ and $\mu_R$ are the factorization and renormalization scales as in the $gg \to t\bar{b}H^{\pm}$ process, and the kinematical variables are $\tau = x_1 x_2$, $x_{1,2} = \sqrt{\tau}e^{\pm y^*}$, $\hat{s} = \tau s$. $\hat{\theta}$ is the polar angle of the $t$-quark in the CM system of the $g\bar{b} \to tH^{\pm}$ scattering, and $\beta_{34} = \hat{s}^{-1}\sqrt{(\hat{s} - m_t^2 - m_{H^{\pm}}^2)^2 - 4m_t^2 m_{H^{\pm}}^2}$. $Q^2$ is the virtuality of the incoming $b$-quark and $z$ is identified with the ratio of the CM energies of the $gb$ system and the $gg$ system.

Note that since the double-counting contribution should be subtracted from the sum of the positive processes, this weight is negative for double-counting events. This means that if all three processes are run simultaneously in PYTHIA, the total cross-section will be correctly matched.

With use of MATCHIG, issues of double-counting in our event generator are resolved. However, we shall not use the Monte-Carlo event generator, PYTHIA, to calculate the precise normalisation of the cross-sections, for though it gives an accurate description of the simulated data in both the low and high transverse momenta regions (with the inclusion of the external process MATCHIG), we can more accurately determine these by taking the leading order cross-section multiplied by an appropriate $k$-factor. The reason for this is that the matched sum is still normalised to the LO total cross-section, we renormalise it to NLO precision using CTEQ6M parton densities and the corresponding value of $\lambda_{\overline{MS}}^{n_f=5} = 226$ MeV in the computations given in Ref.[25, 26], which has been shown to be in good agreement with the one performed in Ref.[27]. For a Higgs boson mass of 300 GeV and in the $\tan\beta$ region of 30–50 considered here, the correction varies very little and can be well approximated with a constant factor of 1.2.

### 3.2 Simulations of the $H^\pm \to \tau\nu$ decay mode

As has already been mentioned, the $\tau\nu$ decay channel offers a high transverse momenta, $p_T$, of the $\tau$ and a large missing energy signature that can be discovered at the LHC over a vast region of the parameter space, where constraints have already been determined [17, 18]. To simulate this the events were generated in PYTHIA using the $gb \to tH^\pm$ process, explicitly using the mechanism $pp \to t(b)H^\pm \to jjb(b)\tau\nu$. That is, the associated top quark is required to decay hadronically, $t \to jjb$. The charged Higgs decays into a $\tau$ lepton, $H^\pm \to \tau^\pm\nu_\tau$, and the hadronic decays of the $\tau$ are considered. The backgrounds considered are QCD, $W+$ jets, single top production $Wt$, and $t\bar{t}$, with one $W \to jj$ and the other $W^\pm \to \tau^\pm\nu_\tau$.

The width of the process $H^\pm \to \tau^\pm\nu_\tau$ is:

$$\Gamma(H^- \to \tau^-\nu_\tau) \simeq \frac{m_{H^\pm}}{8\pi v^2}\left[m_\tau^2 \tan^2\beta\left(1 - \frac{m_\tau^2}{m_{H^\pm}^2}\right)\right]\left(1 - \frac{m_\tau^2}{m_{H^\pm}^2}\right). \quad (25)$$

If the decay $H^\pm \to tb$ is kinematically allowed, comparing its width with Eq.(25) can give a rough estimate of the $H^\pm \to \tau^\pm\nu_\tau$ branching ratio:

$$\begin{aligned} Br(H^\pm \to \tau^\pm\nu_\tau) &\simeq \frac{\Gamma(H^\pm \to \tau^\pm\nu_\tau)}{\Gamma(H^\pm \to tb) + \Gamma(H^\pm \to \tau^\pm\nu_\tau)} \\ &= \frac{m_\tau^2 \tan^2\beta}{3(R_t^{-1})^2(m_t^2\cot^2\beta + m_b^2\tan^2\beta) + m_\tau^2\tan^2\beta}. \end{aligned} \quad (26)$$

Note that a measurement of the signal rate in $H^\pm \to \tau^\pm\nu_\tau$ can allow a determination of $\tan\beta$.

Our approach for this process is as follows:

- We first searched for events having one $\tau$ jet, two light non-$\tau$ jets and at least one (or two) $b$-jets. There is no isolated hard lepton in this configuration.
- A $W$-boson from the top quark decay was first reconstructed using a light jet pair. Note that we retained all the combinations of light jets that satisfy $|m_{jj} - m_W|^2 < 25$GeV. We then rescaled the four momenta of such jets in order to arrive at the correct $W$-boson mass.
- We then reconstructed the top quark by pairing the above constructed $W$-boson with the bottom quarks. Choosing the combination which minimises $\chi^2 = (m_{jjb} - m_t)^2$, we only retained the events that satisfied $|m_{jjb} - m_t| < 25$GeV.
- In this case, due to the presence of missing energy (the neutrino) in the charged Higgs decay, we can not reconstruct the charged Higgs mass. Instead we constructed the transverse mass of the charged Higgs.

Note that we were required to impose additional cuts, namely:

- $\mathcal{N}_1$: On the transverse momenta, $p_T > 100$GeV. A hard cut that allows events for a more massive charged Higgs bosons to pass through. This cut is satisfied by the events that originate from $W$ with large $p_T$. This cut is severe for relatively light charged Higgs bosons (up to 200GeV) as it removes a large number of events, but is a very good cut for a relatively heavy Higgs.
- $\mathcal{N}_2$: On the missing transverse momenta, $p_T^{miss} > 100$GeV. Another hard cut which removes any possible QCD backgrounds, as typically QCD events have no hard leptons. Again this cut is problematic for relatively light Higgs masses, as it removes a large number of events.
- $\mathcal{N}_3$: Finally, a cut on the azimuthal angle between $p_T$ and $p_T^{miss}$ was made. This cut removes the events coming from $W$ with large $p_T$. The decay product of such high $p_T$ $W$-bosons will satisfy the cuts on $p_T^{\tau}$ and $p_T^{miss}$ as defined above. Such events originating from large $p_T$ $W$-bosons gives a large boost to the final products, and hence forces a rather small opening in the angle between the $\tau$ and $\nu$. In the case of the charged Higgs (whose mass is much greater than the $W$'s) the boost is relatively smaller, and this gives a relatively large angle between the $\tau$ and $\nu$. As such we cut the azimuthal angle for $\delta\phi > 1$ rad. This cut becomes much more effective as we move to larger Higgs masses, as the Lorentz boost for larger masses is much less, and hence there shall be larger angles between the final products.

Note also, that in order to add a greater degree of realism to our analysis we have also required that the:

- $B$-tagging efficiency be 60%.
- $c$-jets being misidentified as $b$-jets at 10%.
- light jets be misidentified as $b$-jets at 3%.
- $\tau$ jet tagging efficiency be 70%,

which is somewhat more optimistic than current ATLAS results [17].

In Fig.5 we have plotted the transverse mass of the charged Higgs in the $H \to \tau\nu$ decay for a luminosity of $300fb^{-1}$, scaled to $30fb^{-1}$. In the plot the three lines correspond to positive events (where all three subprocesses are considered together), negative events (the amount to be subtracted to avoid double-counting) and the final matched events. The three panels correspond to different values of $R^{-1}$, as indicated. From this it can be observed that the resonance just below 250GeV is not particularly sensitive to the value of $R^{-1}$, the height of peak is slightly larger for higher values of $R^{-1}$. To further demonstrate the value of this process, we present in table 1 a comparison of the number of signal to background events, where the uncertainty in cross-section measurements is estimated as [10]:

$$\frac{\triangle(\sigma \times BR)}{(\sigma \times BR)} = \sqrt{\frac{S+B}{S^2}},$$

where $S$ and $B$ are signal and background events respectively.

The numerical results of our analysis are therefore summarized in table 1. The table shows that for a reasonable range of input parameters the cross-sections at the LHC can be measured with a 10% accuracy for a luminosity of $\mathcal{L}$ = 100 fb$^{-1}$, whereas the measurement can be improved substantially for higher luminosities. Note that the error in the measurement of $\tan\beta$ is consistent with the observations made in Ref.[10]. For our analysis we have taken the

| | $R_{33}^{-1} = 0.7$ | $R_{33}^{-1} = 1$ | $R_{33}^{-1} = 1.3$ |
|---|---|---|---|
| $\sigma$ (fb) | 204 | 249 | 273 |
| Pre-selection | $48 \times 10^{-3}$ | $48 \times 10^{-3}$ | $48 \times 10^{-3}$ |
| $\mathcal{N}_1$ | $12.8 \times 10^{-3}$ | $13 \times 10^{-3}$ | $13 \times 10^{-3}$ |
| $\mathcal{N}_2$ | $61 \times 10^{-4}$ | $67 \times 10^{-4}$ | $66 \times 10^{-4}$ |
| $\mathcal{N}_3$ | $47 \times 10^{-4}$ | $53 \times 10^{-4}$ | $52 \times 10^{-4}$ |
| $\triangle\,(\sigma \times \text{BR})\,/\,(\sigma \times \text{BR})\,(\mathcal{L} = 100 fb^{-1})$ | 10.6 % | 9.5 % | 8.6 % |
| $\triangle\,(\sigma \times \text{BR})\,/\,(\sigma \times \text{BR})\,(\mathcal{L} = 300 fb^{-1})$ | 6.2 % | 5.5 % | 5 % |

Table 1. Cumulative efficiencies of cuts and estimated errors for measurements of a signal cross-section for the process $pp \to t(b)H(\to \tau^{had}\nu)$. For these numbers we have fixed $m_{H^\pm} = 300$ GeV.

error in the measurement of the cross-section in this channel to be 10% for a luminosity of 100 fb$^{-1}$ and 7.5% for a luminosity of 300 fb$^{-1}$. At this point we would like to note that for our results we have used fast detector simulator ATLFAST [28] and have followed the methodology as given in Ref.[10].

### 3.3 Simulations of the $H^\pm \to tb$ decay mode

Finally, for the decay chain $H^\pm \to tb$, recall that the interaction term of the charged Higgs with the $t$ and $b$ quarks in the 2HDM of type II, as given by Ref.[10], is:

$$\mathcal{L} = \frac{g(R_{33}^{-1})^{-1}}{2\sqrt{2}\, m_W} V_{tb} H^+ \bar{t}\,(m_t \cot\beta(1-\gamma_5) + m_b \tan\beta(1+\gamma_5))\, b + h.c.\,. \tag{27}$$

For the hadroproduction process $gb \to tH^\pm$ (see Fig.3) with the decay mechanism $H^\pm \to tb$, the cross section for $gb \to tH^\pm$ can be written as:

$$\sigma(gb \to tH^\pm) \propto (R_{33}^{-1})^{-2} \left( m_t^2 \cot^2\beta + m_b^2 \tan^2\beta \right)\,. \tag{28}$$

Therefore, the decay width of $H^- \to \bar{t}b$ is given by:

$$\Gamma(H^- \to \bar{t}b) \simeq \frac{3\, m_{H^\pm} (R_{33}^{-1})^{-2}}{8\,\pi v^2} \left[ \left( m_t^2 \cot^2\beta + m_b^2 \tan^2\beta \right) \left( 1 - \frac{m_t^2}{m_{H^\pm}^2} - \frac{m_b^2}{m_{H^\pm}^2} \right) - \frac{4 m_t^2 m_b^2}{m_{H^\pm}^2} \right]$$
$$\times \left[ 1 - \left( \frac{m_t + m_b}{m_{H^\pm}} \right)^2 \right]^{1/2} \left[ 1 - \left( \frac{m_t - m_b}{m_{H^\pm}} \right)^2 \right]^{1/2}, \tag{29}$$

where the factor 3 takes into account the number of colours. The final state of the hadroproduction process contains two top quarks, one of which we required to decay semi-leptonically to provide the trigger, $t \to \ell\nu b$ ($\ell = e, \mu$), and the other hadronically, $\bar{t} \to jjb$. The main background comes from $t\bar{t}b$ and $t\bar{t}q$ production with $t\bar{t} \to WbWb \to \ell\nu bjjb$.

As such, we have used the production channel $pp \to tH^\pm$ for this decay, and have tried to reconstruct the charged Higgs mass. That is, we have the following decay chain:

$$pp \to tH^\pm \to t(tb) \to (\ell\nu_\ell b)(jjb)b \to \ell jjbbb\nu\,. \tag{30}$$

The procedure we have used in reconstructing the masses is:

- We initially searched for one isolated lepton (both electrons and muons) with at least three tagged $b$-jets (this is done in order to include processes like $gg \to tbH$) and at least two non $b$-jets. Furthermore, we used the cuts, where for $b$ and non-$b$ jets we used the same $p_T$ cuts, $p_T^e > 20$GeV, $p_T^\mu > 6$GeV, $p_T^j > 30$GeV and $|\eta| < 2.5$.
- Next we tried to reconstruct the $W$ mass (where the $W$ originates from the top decay) in both leptonic ($W \to \ell \nu$) and hadronic ($W \to jj$) decays. For the leptonic decay we attributed the missing $p_T$ to the emergence of neutrinos from the leptonic $W$ decay. Using the actual $W$ mass we then reconstructed the longitudinal neutrino momentum. This gives a two fold ambiguity, both corresponding to the actual $W$ mass, and neglecting the event if it gives an unphysical solution. Choosing both solutions the second $W$ is reconstructed in the jet mode. We constructed all possible combinations of non-$b$ jets and have plotted the invariant mass of the jets ($m_{jj}$), retaining only those combinations of jets which are consistent with $|m_{jj} - m_W| < 10$GeV. Note that the rescaling is done by scaling the four momenta of the jets with the $W$ mass, that is, $p'_j = p_j \times m_W / m_{jj}$.
- We then attempted to reconstruct the top quarks, where we have, at present, reconstructed two $W$ bosons and three tagged $b$ jets. There can be six different combinations of $W$'s and $b$-jets that can give top quarks. As such, we chose the top quarks which minimise

$$(m_{jjb} - m_t)^2 + (m_{\ell \nu b} - m_t)^2 .$$

- Finally, we retained the top quarks that satisfy $|m_{jjb} - m_t| < 12$GeV and $|m_{\ell\nu b} - m_t| < 12$GeV. This leaves two top quarks and one $b$-jet. There can be two possible combinations, where we retained both. It should be noted that only one of the combinations is the true combination (the combination that emerged from a charged Higgs), the other combination being combinatorial backgrounds.

Using these techniques we can now generate the correlation plot of the two LHC processes considered here, see Fig.6. In these plots we have considered three different values of $R_{33}^{-1}$, where these lines of constant $R_{33}^{-1}$ are generated from three values of $\tan\beta$ (that is, $\tan\beta = 30, 40$ and 50). Note that though this mode has a much larger branching ratio than $H^\pm \to \tau\nu$, it has at least three $b$-jets in its final state. As such, the combinatorial backgrounds associated with this channel make it a challenging task to work with [10], and not the best discovery channel for a charged Higgs at the LHC.

## 4. Charged Higgs at $B$-factories

Having now reviewed how a massive charged Higgs may be detected at the LHC, we shall now place greater constraints on the charged Higgs parameters by utilising the successful $B$ factory results from KEK and SLAC. Note that $B$ physics shall be a particularly fertile ground to place constraints on a charged Higgs. For example, it is well known that limits from $b \to s\gamma$ can give stronger constraints in generic 2HDMs than in SUSY models [29]!

The $B$ decays of most interest here are those including a final $\tau$ particle, namely $B \to D\tau\nu$ and $B \to \tau\nu$[8]. An important feature of these processes is that a charged Higgs boson can contribute to the decay amplitude at tree-level in models such as the 2HDM and the MSSM. From the experimental perspective, since at least two neutrinos are present in the final state (on the signal side), a full-reconstruction is required for the $B$ decay on the opposite side. For the $B \to D\tau\nu$ process, the branching fraction has been measured at BaBar with

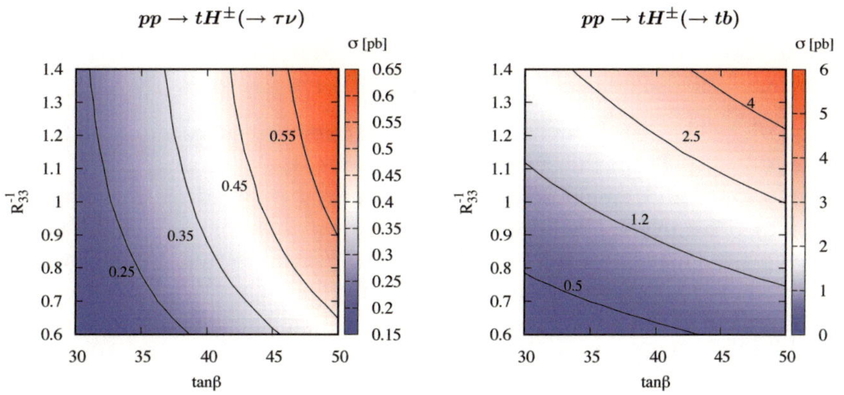


Fig. 6. Contour plots of the cross-sections for the processes $pp \to tH^{\pm}(\to \tau\nu)$ (left) and $pp \to tH^{\pm}(\to tb)$ (right) versus $R_{33}^{-1}$ and $\tan\beta$ with fixed $m_{H^{\pm}}$ =300 GeV [11].

$Br(B \to D^{+}\tau^{-}\bar{\nu}_{\tau}) = 0.86 \pm 0.24 \pm 0.11 \pm 0.06\%$ [30], which is consistent, within experimental uncertainties, with the SM, and with Belle [31]. Note also that the inclusive $b \to c\tau\nu$ branching ratio was determined at the LEP experiments [32]. The $B \to \tau\nu$ process has a smaller branching ratio, as measured by Belle at $(1.79\ ^{+0.56}_{-0.49}\ (stat)\ ^{+0.46}_{-0.51}\ (syst)) \times 10^{-4}$ [33], and at BaBar $(1.2 \pm 0.4 \pm 0.3 \pm 0.2) \times 10^{-4}$ [34] (giving an average of $(1.41\ ^{+0.43}_{-0.42}) \times 10^{-4}$ [35]). Note that the SM predicts $Br(B \to \tau\nu) = (7.57\ ^{+0.98}_{-0.61}) \times 10^{-5}$, where theoretical uncertainties came from $f_B$, the $B$ meson decay constant, which from lattice QCD is $f_B = 191 \pm 13$ MeV. As such, the measurement of these processes will be important targets in coming $B$ factory experiments.

In order to test for the charged Higgs fermion couplings, we now determine the charged Higgs contributions to tauonic $B$ decays, where it is straightforward to write down the amplitudes for the $B \to D\tau\nu$ ($B^{-} \to \overline{D}^{0}\tau^{-}\overline{\nu}$ or $\overline{B}^{0} \to D^{+}\tau^{-}\overline{\nu}$) and $B \to \tau\nu$ processes. We should first like to note that the higgsino diagram contributions, see Fig.1(a), to the $R_{22}^{-1}$ are proportional to square of the charm Yukawa couplings, and since the branching ratio can change only by at most a few percent, we shall neglect such contributions here. Also, as we shall work with large $\tan\beta$ values, $\cot\beta$ terms can be neglected in the Lagrangian.

We can now calculate the charged Higgs effect on the $B \to D\tau\nu$ branching ratio, by utilising the vector and scalar form factors of the $B \to D$ transition. These are obtained using the effective Lagrangian for $b \to c\tau\nu$ operators as given by

$$\mathcal{L}_{\rm eff} = -\frac{G_{\rm F}}{\sqrt{2}} V_{cb}\bar{c}\gamma_{\mu}(1-\gamma_5)b\bar{\tau}\gamma^{\mu}(1-\gamma_5)\nu_{\tau} + G_{\rm S}\bar{c}b\bar{\tau}(1-\gamma_5)\nu_{\tau} + G_{\rm P}\bar{c}\gamma_5 b\bar{\tau}(1-\gamma_5)\nu_{\tau} + \text{h.c.}\,, \quad (31)$$

where $G_{\rm S}$ and $G_{\rm P}$ are scalar and pseudo-scalar effective couplings. These couplings are given from Eqs.(10), (11) and the similarly derived effective Lagrangian for charged leptons:

$$G_{\rm S} \equiv \frac{\tan^2\beta M_{\tau}}{2v^2 M_{H^{\pm}}^2}[\hat{R}_e^{-1}]_{33}(M_b[\hat{R}_d^{-1}]_{22}V_{cb} + M_c V_{cb}\cot^2\beta)\,, \quad (32)$$

$$G_{\rm P} \equiv \frac{\tan^2\beta M_{\tau}}{2v^2 M_{H^{\pm}}^2}[\hat{R}_e^{-1}]_{33}(M_b[\hat{R}_d^{-1}]_{22}V_{cb} - M_c V_{cb}\cot^2\beta)\,, \quad (33)$$

where we shall now omit a prime from the fields in mass eigenstates. Recall that we shall neglect higgsino diagram contributions to the $[\hat{R}_d^{-1}]_{22}$ proportional to the square of the charm Yukawa couplings, and also neglect the last terms in $G_S$ and $G_P$.

In the heavy quark limit, these form factors can be parameterized by a unique function called the Isgur-Wise function. From the semi-leptonic decays $B \to Dl\nu$ and $B \to D^* l\nu$ $(l = e, \mu)$, the Isgur-Wise function is obtained in a one-parameter form, including the short distance and $1/M_Q$ $(Q = b, c)$ corrections. The short distance corrections for $B \to D\tau\nu$ have also been calculated previously [36]. Here we adopt this Isgur-Wise function, but do not include the short distance and the $1/M_Q$ corrections for simplicity.

Using the definitions,

$$x \equiv \frac{2p_B \cdot p_D}{p_B^2}\ ,\ y \equiv \frac{2p_B \cdot p_\text{ø}}{p_B^2}\ ,\ r_D \equiv \frac{M_D^2}{M_B^2}\ ,\ r_\text{ø} \equiv \frac{M_\text{ø}^2}{M_B^2}\ , \tag{34}$$

the differential decay width is given by

$$\frac{d^2\Gamma[B \to D\tau\nu]}{dxdy} = \frac{G_F^2|V_{cb}|^2}{128\pi^3} M_B^5 \rho_D(x,y)\ , \tag{35}$$

where

$$\begin{aligned}
\rho_D(x,y) &\equiv [|f_+|^2 g_1(x,y) + 2\mathrm{Re}(f_+ f_-'^*) g_2(x,y) + |f_-'|^2 g_3(x)]\ ,\\
g_1(x,y) &\equiv (3 - x - 2y - r_D + r_\text{ø})(x + 2y - 1 - r_D - r_\text{ø}) - (1 + x + r_D)(1 + r_D - r_\text{ø} - x)\ ,\\
g_2(x,y) &\equiv r_\text{ø}(3 - x - 2y - r_D + r_\text{ø})\ ,\\
g_3(x) &\equiv r_\text{ø}(1 + r_D - r_\text{ø} - x)\ ,\\
f_-' &\equiv \{f_- - \Delta_S[f_+(1 - r_D) + f_-(1 + r_D - x)]\}\ ,\\
f_\pm &= \pm\frac{1 \pm \sqrt{r_D}}{2\sqrt[4]{r_D}}\xi(w),\ \ (w = \frac{x}{2\sqrt{r_D}})\ .
\end{aligned}$$

Here $\Delta_S \equiv \frac{\sqrt{2}G_S M_B^2}{G_F V_{cb} M_\tau (M_b - M_c)}$. We use the following form of the Isgur-Wise function.

$$\begin{aligned}
\xi(w) &= 1 - 8\rho_1^2 z + (51\rho_1^2 - 10)z^2 - (252\rho_1^2 - 84)z^3\ ,\\
z &= \frac{\sqrt{w+1} - \sqrt{2}}{\sqrt{w+1} + \sqrt{2}}\ .
\end{aligned}$$

For the slope parameter we use $\rho_1^2 = 1.33 \pm 0.22$ [36, 37].

Similarly, for the $B \to \tau\nu$ process, the relevant four fermion interactions are those of the $b \to u\tau\nu$ type [8]:

$$\mathcal{L}'_{\text{eff}} = -\frac{G_F}{\sqrt{2}} V_{ub}\bar{u}\gamma_\mu(1-\gamma_5)b\bar{\tau}\gamma^\mu(1-\gamma_5)\nu_\tau + G_S'\bar{u}b\bar{\tau}(1-\gamma_5)\nu_\tau + G_P'\bar{u}\gamma_5 b\bar{\tau}(1-\gamma_5)\nu_\tau + \text{h.c.}\ , \tag{36}$$

$$G_S' \equiv \frac{\tan^2\beta M_\tau}{2v^2 M_{H^\pm}^2}[\hat{R}_e^{-1}]_{33}(M_b[\hat{R}_d^{-1}]_{11} V_{ub} + M_u V_{ub}\cot^2\beta)\ , \tag{37}$$

$$G_P' \equiv \frac{\tan^2\beta M_\tau}{2v^2 M_{H^\pm}^2}[\hat{R}_e^{-1}]_{33}(M_b[\hat{R}_d^{-1}]_{11} V_{ub} - M_u V_{ub}\cot^2\beta)\ . \tag{38}$$

Using the matrix elements

$$\langle 0|\overline{u}\gamma^{\mu}\gamma_5 b|B^-\rangle = if_{\rm B}p^{\mu}\,,$$
$$\langle 0|\overline{u}\gamma_5 b|B^-\rangle = -if_{\rm B}\frac{M_{\rm B}^2}{M_b}\,,$$

the decay width of $B \to \tau\nu$ in the SM is given by:

$$\Gamma[B \to \tau\nu_\tau]_{SM} = \frac{G_F^2}{8\pi}|V_{ub}|^2 f_B^2 m_\tau^2 m_B \left(1 - \frac{m_\tau^2}{m_B^2}\right)^2, \tag{39}$$

which in the presence of a charged Higgs boson, is modified by a multiplicative factor to:

$$\Gamma[B \to \tau\nu_\tau]_{2HDM} = \Gamma[B \to \tau\nu_\tau]_{SM} \times \left(1 - \frac{m_B^2}{m_{H^\pm}^2}\tan^2\beta\right)^2, \tag{40}$$

in the effective limits we have adopted. Note that our input parameters are the projected values for Super$B$, that is, we shall use $f_{\rm B} = 200 \pm 30$MeV in our numerics.

Note that this link can be understood by recalling that in our generalized case of MFV, that is Eq.(16), the scalar and pseudo-scalar couplings, Eqs.(32), (33), (37), and (38) can be obtained by the following replacement.

$$[\hat{R}_d^{-1}]_{22} \to \frac{1}{1 + [E_{\tilde{g}}^{(3)} - E_{\tilde{g}}'^{(32)}]\tan\beta}, \tag{41}$$

$$[\hat{R}_d^{-1}]_{11} \to \frac{1}{1 + [E_{\tilde{g}}^{(3)} - E_{\tilde{g}}'^{(31)}]\tan\beta}, \tag{42}$$

where $E_{\tilde{g}}^{(i)}$ and $E_{\tilde{g}}'^{(ij)}$ were defined in section 2.1. Notice that the right-handed sides of the above equations are approximately the same because $E_{\tilde{g}}'^{(31)} \approx E_{\tilde{g}}'^{(32)}$. This is the generalization of $[\hat{R}_d^{-1}]_{11} \approx [\hat{R}_d^{-1}]_{22}$, which follows from fact that the higgsino diagram contribution can be neglected in the evaluation with the $[\hat{R}_d^{-1}]_{11}$ and $[\hat{R}_d^{-1}]_{22}$.

Using these results we have generated the contour plots in Fig.7(b), where a correlation of the $B \to D\tau\nu$ ($B^- \to \overline{D}^0\tau^-\overline{\nu}$ or $\overline{B}^0 \to D^+\tau^-\overline{\nu}$) and $B \to \tau\nu$ branching ratios for various values of $\tan\beta$ and $\widehat{R}_d^{-1}$ and $m_{H^\pm} = 300$GeV, have been given.

## 5. Determination of the effective couplings

Collecting our numerical results from section 3 and the branching ratios calculated in the previous subsection, we have generated the plots in Figs.7 and 8. In these figures we can see correlations of the LHC cross-sections with the two $B$ processes, where in these plots we have varied $\tan\beta$ in the range $30 < \tan\beta < 50$ for different values of $R_{ii}^{-1}$ ($ii = 11, 33$). Fig.7(a) shows the correlation of the LHC observables, whilst the correlation of $B$-decay branching ratios in Fig.7(b) gives the same line for different values of $R_{11}^{-1}$. The reason for this can be seen from Eq.(36) where $R_{ii}^{-1}$ and $\tan\beta$ arise from the same combination ($\equiv R_{ii}^{-1}\tan^2\beta$) in

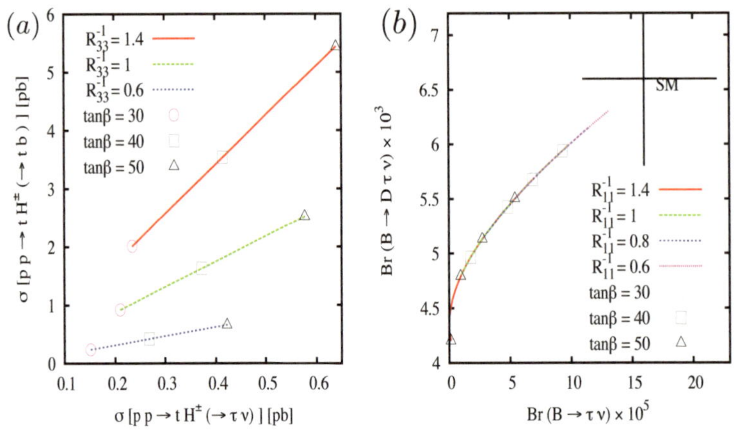


Fig. 7. Correlation plots of the cross-sections for the processes $pp \to t(b)H^{\pm}(\to \tau\nu)$ and $pp \to t(b)H^{\pm}(\to tb)$ for three values of $R_{33}^{-1}$ and $\tan\beta$ (left) and of the branching ratios for $B \to D\tau\nu$ and $B \to \tau\nu$ (right) for various values of $\tan\beta$ and $\widehat{R}_d^{-1}$ with fixed $m_{H^{\pm}} = 300$ GeV [11].

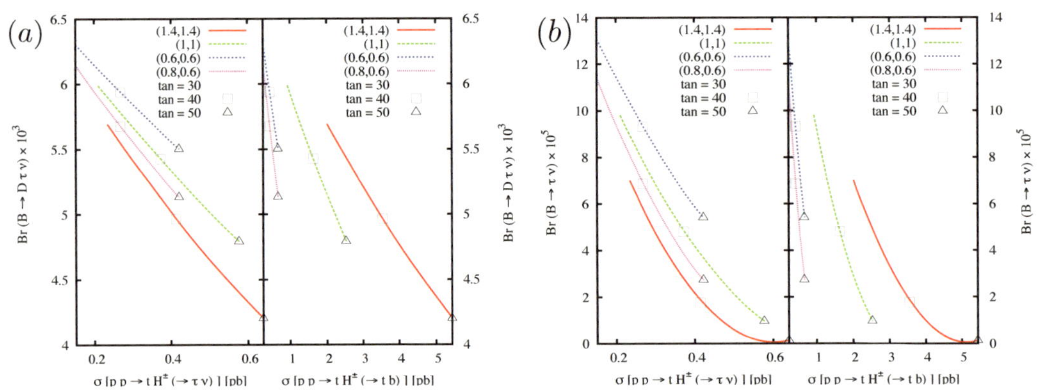


Fig. 8. Contour plots of the $B \to D\tau\nu$ branching ratio correlated with the cross-section $\sigma(pp \to t(b)H^{\pm}(\to \tau\nu)$ (*a*) left) and $\sigma(pp \to t(b)H^{\pm}(\to tb)$ (*a*) right), and the $B \to \tau\nu$ branching ratio correlated with the cross-section $\sigma(pp \to t(b)H^{\pm}(\to \tau\nu)$ (*b*) left) and $\sigma(pp \to t(b)H^{\pm}(\to tb)$ (*b*) right), for various values of $\tan\beta$ and $R^{-1}$ (the bracketed numbers in the key refer to the appropriate $R^{-1}$ for each process being considered)[11].

the tauonic $B$-decays considered in this work. Hence the measurement of these two $B$-decays will only give an estimate of the product of $R_{11}^{-1}$ and $\tan\beta$. However, by considering the correlations of the $B$-decay observables with LHC observables, as shown in Fig.8, one can remove this degeneracy. So in principle it is possible to measure the four parameters ($\tan\beta$ and $R_{ii}^{-1}$ with $ii = 11, 22, 33$) using the six correlation plots shown in Figs.7 and 8.

The primary question to be answered in this effective test of the charged Higgs couplings is "to what precision can we test $R^{-1}$?". From our simulations we can safely assume that the LHC shall determine, to some level of precision, values for $m_{H\pm}$ and/or $\tan\beta$. These values can then be converted into a value for $R^{-1}$ with all the precision afforded to us from the results of the $B$-factory experiments, as demonstrated pictorially in Fig.8. Assuming the charged Higgs boson mass to be known (taken to be 300 GeV in our present analysis) we have obtained cross-section measurement uncertainties as given in table 1. As can be seen from this, it might be possible to measure $R_{33}^{-1}$ and $\tan\beta$ with an accuracy of about 10% at high luminosity. Armed with this information about $\tan\beta$, from the LHC measurements, it can then be taken as an input to the $B$-decay measurements, namely $B \to \tau\nu$ and $B \to D\tau\nu$. In Ref.[10] it was inferred that for large values of $\tan\beta$ ($\geq 40$), measurements to a precision of 6-7% for high luminosity LHC results are possible. Our results are consistent with these observations. Future Super-$B$ factories are expected to measure the $B \to \tau\nu$ and $B \to D\tau\nu$ to a precision of 4% and 2.5% respectively [38]. The present world average experimental results for tauonic $B$-decays are $BR(B \to \tau\nu) = (1.51 \pm 0.33) \times 10^{-4}$ and $BR(B \to D\tau\nu)/BR(B \to D\mu\nu) = (41.6 \pm 11.7 \pm 5.2)\%$ [30, 38]. Presently if one uses UTfit prescription of $|V_{ub}|$ then there is substantial disagreement between experimental and SM estimates for the branching fractions of $B \to \tau\nu$. Recently, proposals have been given in Ref.[39] to reduce this tension between experimental and theoretical SM values of $B \to \tau\nu$. Transforming the improved projected theoretical information of these decays along with future Super-$B$ factory measurements one can measure $R_{11}^{-1}$ and $R_{22}^{-1}$ to a fairly high precision.

To summarise, we have tried to demonstrate that at the LHC alone it is possible to measure the charged Higgs boson couplings, namely $\tan\beta$ and $R_{33}^{-1}$, to an accuracy of less than 10%. Combining this information from the LHC with improved $B$-factory measurements, one can measure all four observables indicated in the introduction. These observables represent effective couplings of a charged Higgs boson to the bottom quark and the three generations of up-type quarks, thus demonstrating that it is possible to test the charged Higgs boson couplings to quarks by the combination of low energy measurements at future Super-$B$ factories and charged Higgs boson production at the LHC. Something which shall be realisable in the very near future as results from the LHC are already starting to emerge [17, 18], and which will require more refined analyses in the near future.

## 6. References


[1] P. Fayet and S. Ferrara, Phys. Rept. 32, 249 (1977); H.P. Nilles, Phys. Rep. 110, 1 (1984); H.E. Haber and G. L. Kane, Phys. Rep. 117, 75 (1985).

[2] S.L. Glashow and S. Weinberg, Phys. Rev. D15, 1958 (1977).

[3] H. Georgi and D.V. Nanopoulos, Phys. Lett. 82B, 95 (1979).

[4] H.E. Haber, G.L. Kane and T. Sterling, Nucl. Phys. B161, 493 (1979).

[5] L.J. Hall and M.B. Wise, Nucl. Phys. B187, 397 (1981).

[6] J.F. Donoghue and L.F. Li, Phys. Rev. D19, 945 (1979).

[7] W.S. Hou, Phys. Lett. B296, 179 (1992); D. Chang, W.S. Hou and W.Y. Keung, Phys. Rev. D48, 217 (1993); D. Atwood, L. Reina and A. Soni, Phys. Rev. D55, 3156 (1997).

[8] H. Itoh, S. Komine and Y. Okada, Prog. Theor. Phys. 114, 179 (2005) [arXiv:hep-ph/0409228].

[9] L. J. Hall, R. Rattazzi and U. Sarid, Phys. Rev. D 50, 7048 (1994).

[10] K. A. Assamagan, Y. Coadou and A. Deandrea, Eur. Phys. J. directC 4, 9 (2002) [arXiv:hep-ph/0203121].

[11] A. S. Cornell, A. Deandrea, N. Gaur, H. Itoh, M. Klasen, Y. Okada, Phys. Rev. D81, 115008 (2010).

[12] M. Gorbahn, S. Jäger, U. Nierste and S. Trine, arXiv:0901.2065 [hep-ph].

[13] T. Blazek, S. Raby and S. Pokorski, Phys. Rev. D 52, 4151 (1995).
[14] K. S. Babu and C. F. Kolda, Phys. Rev. Lett. 84, 228 (2000);
[15] [LEP Higgs Working Group for Higgs boson searches and ALEPH Collaboration], arXiv:hep-ex/0107031.
[16] V. M. Abazov *et al.* [D0 Collaboration], Phys. Lett. B 682, 278 (2009) [arXiv:0908.1811 [hep-ex]].
[17] The ATLAS Collaboration, ATLAS-CONF-2011-138 August 31, 2011.
[18] The CMS Collaboration, CMS PAS HIG-11-008 July 22, 2011.
[19] J. F. Gunion, H. E. Haber, F. E. Paige, W. K. Tung and S. S. Willenbrock, Nucl. Phys. B 294, 621 (1987); J. L. Diaz-Cruz and O. A. Sampayo, Phys. Rev. D 50, 6820 (1994).
[20] A. Krause, T. Plehn, M. Spira and P. M. Zerwas, Nucl. Phys. B 519, 85 (1998) [arXiv:hep-ph/9707430]; Y. Jiang, W. g. Ma, L. Han, M. Han and Z. h. Yu, J. Phys. G 24, 83 (1998) [arXiv:hep-ph/9708421]; A. A. Barrientos Bendezu and B. A. Kniehl, Nucl. Phys. B 568, 305 (2000) [arXiv:hep-ph/9908385]; O. Brein and W. Hollik, Eur. Phys. J. C 13, 175 (2000) [arXiv:hep-ph/9908529].
[21] D. A. Dicus, J. L. Hewett, C. Kao and T. G. Rizzo, Phys. Rev. D 40, 787 (1989); A. A. Barrientos Bendezu and B. A. Kniehl, Phys. Rev. D 59, 015009 (1999) [arXiv:hep-ph/9807480]; Phys. Rev. D 61, 097701 (2000) [arXiv:hep-ph/9909502]; Phys. Rev. D 63, 015009 (2001) [arXiv:hep-ph/0007336]; O. Brein, W. Hollik and S. Kanemura, Phys. Rev. D 63, 095001 (2001) [arXiv:hep-ph/0008308].
[22] J. Alwall, arXiv:hep-ph/0503124.
[23] T. Sjöstrand, S. Mrenna and P. Skands, JHEP 0605, 026 (2006).
[24] J. Alwall and J. Rathsman, "Improved description of charged Higgs boson production at hadron colliders," JHEP 0412 (2004) 050 [arXiv:hep-ph/0409094].
[25] T. Plehn, Phys. Rev. D 67, 014018 (2003).
[26] E. L. Berger, T. Han, J. Jiang and T. Plehn, Phys. Rev. D 71, 115012 (2005).
[27] S. H. Zhu, Phys. Rev. D 67 075006 (2003).
[28] E. Richter-Was *et al.* , *ATLFAST 2.2: A fast simulation package for ATLAS*, ATL-PHYS-98-131.
[29] R. Bose and A. Kundu, arXiv:1108.4667 [hep-ph].
[30] B. Aubert *et al.* [BABAR Collaboration], Phys. Rev. Lett. 100, 021801 (2008) [arXiv:0709.1698 [hep-ex]].
[31] K. Abe *et al.* [Belle Collaboration], Phys. Lett. B 526, 258 (2002) [arXiv:hep-ex/0111082].
[32] G. Abbiendi *et al.* [OPAL Collaboration], Phys. Lett. B 520 (2001), 1; R. Barate *et al.* [ALEPH Collaboration], Eur. Phys. J. C 19 (2001), 213.
[33] K. Ikado *et al.*, Phys. Rev. Lett. 97, 251802 (2006) [arXiv:hep-ex/0604018].
[34] B. Aubert *et al.* [BABAR Collaboration], Phys. Rev. D 77, 011107 (2008) [arXiv:0708.2260 [hep-ex]]; B. Aubert *et al.* [BABAR Collaboration], Phys. Rev. D 76, 052002 (2007) [arXiv:0705.1820 [hep-ex]].
[35] Heavy Flavour Averaging Group, http://www.slac.stanford.edu.au/xorg/hfag/.
[36] T. Miki, T. Miura and M. Tanaka, in Shonan Village 2002, Higher luminosity B factories, arXiv:hep-ph/0210051.
[37] C. G. Boyd, B. Grinstein and R. F. Lebed, Phys. Rev. D 56 (1997), 6895; I. Caprini, L. Lellouch and M. Neubert, Nucl. Phys. B 530 (1998), 153.
[38] T. Browder, M. Ciuchini, T. Gershon, M. Hazumi, T. Hurth, Y. Okada and A. Stocchi, JHEP 0802, 110 (2008);
[39] E. Lunghi and A. Soni, arXiv:0912.0002 [hep-ph];

# 3

# Introduction to Axion Photon Interaction in Particle Physics and Photon Dispersion in Magnetized Media

Avijit K. Ganguly
*Banaras Hindu University (MMV), Varanasi, India*

## 1. Introduction

Symmetries, global or local, always play an important role in the conceptual aspects of physics be in broken or unbroken phase. Spontaneous breaking of the continuous symmetries always generates various excitations with varying mass spectra. Axion is one of that type, generated via spontaneous breaking of a global Chiral U(1) symmetry named after its discoverers, Peccei and Queen. This symmetry is usually denoted by $U(1)_{PQ}$. To give a brief introduction to this particle and its origin we have to turn our attention to the development of the standard model of particle physics and its associated symmetries. The standard model of particle physics describes the strong, weak and electromagnetic interactions among elementary particles. The symmetry group for this model is, $SU_c(3) \times SU(2) \times U(1)$. The strong interaction ( Quantum Chromo Dynamics (QCD)) part of the Lagrangian has SU(3) color symmetry and it is given by,

$$\mathcal{L} = -\frac{1}{2g^2} \mathrm{Tr}\ \mathrm{F}^a_{\mu\nu}\mathrm{F}^{\mu\nu}_a + \bar{q}(i\not{D} - m)q. \tag{1}$$

It was realized long ago that, in the limit of vanishingly small quark masses (chiral limit), Strong interaction lagrangian has a global $U(2)_V \times U(2)_A$ symmetry. This symmetry group would further break spontaneously to produce the hadron multiplets. The vector part of the symmetry breaks to iso-spin times baryon number symmetry given by $U(2)_V = SU(2)_I \times U(1)_B$. In nature baryon number is seen to be conserved and the mass spectra of nucleon and pion multiplets indicate that the isospin part is also conserved approximately.

So one is left with the axial vector symmetry. QCD being a nonabelian gauge it is believed that this theory is confining in the infrared region. The confining property of the theory is likely to generate condensates of antiquark quark pairs. Thus $u$- and $d$ quark condensates would have non-zero vacuum expectation values, i.e.,

$$< 0|\bar{u}(0)u(0)|0 >=< 0|\bar{d}(0)d(0)|0 > \neq 0\,. \tag{2}$$

and they would break the $U(2)_A$ symmetry. Now if the axial symmetry is broken, we would expect nearly four degenerate and massless pseudoscalar mesons. Interestingly enough, out of the four we observe three light pseudoscalar Nambu Goldstone (NG) Bosons in nature, i.e., the pions. They are light, $m_\pi \simeq 0$, but the other one (with approximately same mass) is not to be found. Eta meson though is a pseudoscalar meson, but it has mass much greater than the pion ( $m_\eta \gg m_\pi$). So the presence of another light pseudoscalar meson in the hadronic spectrum, seem to be missing. This is usually referred in the literature [(Steven Weinberg , 1975)] as the $U(1)_A$ problem.

### 1.1 Strong CP problem and neutron dipole moment

Soon after the identification of QCD as the correct theory of strong interaction physics, instanton solutions [(Belavin Polyakov Shvarts and Tyupkin , 1975)] for non-abelian gauge theory was discovered. Subsequently, through his pioneering work, 't Hooft [('t Hooft , 1976a),('t Hooft , 1976b)] established that a $\theta$ term must be added to the QCD Lagrangian. The expression of this piece is,

$$\mathcal{L}_\theta = \theta \frac{g^2}{32\pi^2} F_a^{\mu\nu} \tilde{F}_{a\mu\nu}. \tag{3}$$

But in the presence of this term *the axial symmetry* is no more a realizable symmetry for QCD. This term violates Parity and Time reversal invariance, but conserves charge conjugation invariance, so it violates CP. Such a term if present in the lagrangian would predict neutron electric dipole moment. The observed neutron electric dipole moment [(R. J. Crewther, 1978)] is $|d_n| < 3 \times 10^{-26}$ ecm and that requires the angle $\theta$ to be extremely small [$d_n \simeq e\theta m_q / M_N^2$ indicating [(V. Baluni , 1979; R. J. Crewther et. al. , 1980)] $\theta < 10^{-9}$]. This came to be known as the strong CP problem. In order to overcome this problem, Pecci and Queen subsequently Weinberg and Wilckzek [(R. Peccei and H. Quinn , 1977; S. Weinberg , 1978; F. Wilczek , 1978)] postulated the parameter $\theta$ to be a dynamical field with odd parity arising out of some chiral symmetry breaking taking place at some energy scale $f_{PQ}$. With this identification the $\theta$ term of the QCD Lagrangian now changes to,

$$\mathcal{L}_a = \frac{g^2}{32\pi^2} a F_a^{\mu\nu} \tilde{F}_{a\mu\nu}, \tag{4}$$

where $a$ is the axion field. They[(R. Peccei and H. Quinn , 1977; S. Weinberg , 1978; F. Wilczek , 1978)] also provided an estimate of the mass of this light pseudoscalar boson. Although these ultra light objects were envisioned originally to provide an elegant solution to the strong CP problem [(R. Peccei and H. Quinn , 1977),WW,wilczek] (see (R. Peccei , 1996)] for details) but it was realized later on that their presence may also solve some of the outstanding problems in cosmology, like the dark matter or dark energy problem (related to the closure of the universe). Further more their presence if established, may add a new paradigm to our understanding of the stellar evolution. A detailed discussion on the astrophysical and cosmological aspects of axion physics can be found in [(M.S. Turner , 1990; G. G. Raffelt , 1990; G. G. Raffelt , 1997; G .G .Raffelt , 1996; J. Preskill et al , 1983)]. In all the models of axions, the axion photon coupling is realized through the following term in the Lagrangian,

$$\mathcal{L} = \frac{1}{M} a \, \mathbf{E} \cdot \mathcal{B}. \tag{5}$$

Where $M \propto f_a$ the axion coupling mass scale or the symmetry breaking scale and $a$ stands for the axion field. The original version of Axion model, usually known as Peccei-Queen Weinberg-Wilczek model (PQWW), had a symmetry breaking scale that was close to weak scale, $f_w$. Very soon after its inception, the original model, associated with the spontaneous breakdown of the global PQ symmetry at the Electro Weak scale (EW) $f_w$, was experimentally ruled out. However modified versions of the same with their associated axions are still of interest with the symmetry breaking scale lying between EW scale and $10^{12}$ GeV. Since the axion photon/matter coupling constant, is inversely proportional to the breaking scale of the PQ symmetry, $f_a$ and is much larger than the electroweak scale $f_a \gg f_w$, the resulting axion turns out to be very weakly interacting. And is also very light ($m_a \sim f_a^{-1}$) therefore it is often called "the invisible axion model" [(M.Dine et al. , 1981; J. E. Kim , 1979)]. For very good introduction to this part one may refer to[(R. Peccei , 1996)].

There are various proposals to detect axions in laboratory. One of them is the solar axion experiment. The idea behind this is the following, if axions are produced at the core of the Sun, they should certainly cross earth on it's out ward journey from the Sun. From equation [5], it can be established that in an external magnetic field an axion can oscillate in to a photon and vice versa. Hence if one sets up an external magnetic field in a cavity, an axion would convert itself into a photon inside the cavity.This experiment has been set up in CERN, and is usually referred as CAST experiment[(K. Zioutas et al.,, 2005)]. The conversion rate inside the cavity, would depend on the value of the coupling constant ($\frac{1}{M}$), axion mass and the axion flux. Since inside the sun axions are dominantly produced by Primakoff and compton effects. One can compute the axion flux by calculating the axion production rate via primakoff & compton process using the available temp and density informations inside the sun. Therefore by observing the rate of axion photon conversion in a cavity on can estimate the axion parameters. The study of solar axion puts experimental bound on $M$ to be, $M > 1.7 \times 10^{11}$GeV [(Moriyama et al. , 1985),(Moriyama et al. , 1998b)].

The same can be estimated from astrophysical observations. In this situation, it possible to estimate the rate at which the axions would draw energy away form the steller atmosphere by calculating the axion flux (i.e. is axion luminosity) from the following reactions[7]

$$e^{+} + e^{-} \rightarrow \gamma + a\ , \quad e^{-} + \gamma \rightarrow e^{-} + a \tag{6}$$

&

$$\gamma_{plasmon} \rightarrow \gamma + a\ , \quad \gamma + \gamma \rightarrow a. \tag{7}$$

Axions being weakly interacting particles, would escape the steller atmosphere and the star would lose energy. Thus it would affect the age vs luminosity relation of the star. Comparison of the same with observations yields bounds on e.g., axion mass $m_a$ and $M$. A detailed survey of various astrophysical bounds on the parameters of axion models and constraints on them, can be found in ref. [(G .G .Raffelt , 1996)].

In the astrophysical and cosmological studies, mentioned above, medium and a magnetic field are always present. So it becomes important to seek the modification of the axion coupling to photon, in presence of a medium or magnetic field or both. Particularly in some astrophysical situations where the magnetic component, along with medium (usually referred as magnetized medium) dominates. Examples being, the Active Galactic Nuclei (AGN), Quasars, Supernova, the Coalescing Neutron Stars or Nascent Neutron Stars, Magnetars etc. . The magnetic field strength in these situations vary between, $\mathcal{B} \sim 10^{6} - 10^{17}$ G, where some are significantly above the critical, Schwinger value[( J. Schwinger , 1951)]

$$\mathcal{B}_e = m_e^2/e \simeq 4.41 \times 10^{13} \text{ G} \tag{8}$$

[(M. Ruderman , 1991; Duncan & Thompson , 1992)]. In view of this observation and the possibilities of applications of axion physics to these astrophysical as well as cosmological scenarios, it is pertinent to find out the effect of medium and magnetic field to axion photon coupling.

As we already have noted, the axion physics is sensitive to presence of medium and magnetic field. In most of the astrophysical or cosmological situations these two effects are dominant. In view of this it becomes reasonable to study how matter and magnetic field effect can affect the axion photon vertex. Modification to axion photon vertex in a magnetized media was studied in [(A. K. Ganguly , 2006)]. In this document we would present that work and discuss new correction to $a-\gamma$ vertex in a magnetized media. In the next section that we would focus on axion photon mixing effect with tree level axion photon vertex and show how this effect can change the polarization angle and ellipticity of a propagating plane polarized light beam passing through a magnetic field. After that we would elaborate on how the same predictions would get modified if the same process takes place in a magnetized media. This particular study involves diagonalisation of a $3 \times 3$ matrix, so at the end we have added an appendix showing how to construct the diagonalizing matrix to diagonalize a $3 \times 3$ symmetric matrix.

## 2. The loop induced vertex

The axion-fermion ( lepton in this note ) interaction[1] — with $g'_{af} = \left(X_f m_f / f_a\right)$ the Yukawa coupling constant, $X_f$, the model-dependent factors for the PQ charges for different generations of quarks and leptons [(G .G .Raffelt , 1996)], and fermion mass $m_f$– is given by, [(M.Dine et al. , 1981)],

$$\mathcal{L}_{af} = \frac{g'_{af}}{m_f} \sum_f (\bar{\Psi}_f \gamma_\mu \gamma_5 \Psi_f)\, \partial^\mu a, \tag{2.9}$$

The sum over f, in eqn. [2.9], stands for sum over all the fermions, from each family. Although, in some studies, instead of using [2.9], the following Lagrangian has been employed,

$$\mathcal{L}_{af} = -2ig'_{af} \sum_f (\bar{\Psi}_f \gamma_5 \Psi_f) a, \tag{2.10}$$

but, Raffelt and Seckel [( G. Raffelt , 1988)] has pointed out the correctness of using [2.9]. We for our purpose we will make use of [2.9]. We would like to note that the usual axion photon mixing Lagrangian in an external magnetic field turns out to be,

$$\mathcal{L}_{a\gamma} = -g_{a\gamma\gamma} \frac{e^2}{32\pi^2} a \mathrm{F}\tilde{\mathrm{F}}^{\mathrm{Ext}}. \tag{2.11}$$

In equation [2.11] the axion photon coupling constant is described by,

$$g_{a\gamma\gamma} = \frac{1}{f_a}\left[A^{em}_{PQ} - A^{c}_{PQ} \frac{2(4+z)}{3(1+z)}\right], . \tag{2.12}$$

with $z = \frac{m_u}{m_d}$, where $m_u$ and $m_d$ are the masses of the light quarks. Anomaly factors are given by the following relations, $A^{em}_{PQ} = \mathrm{Tr}(Q_f^2) X_f$ and $\delta_{ab} A^{em}_c = \mathrm{Tr}(\lambda_a \lambda_b X_f)$ (and the trace is over

[1] Some of the issues related axion fermion coupling had been reviewed in [(A. K. Ganguly , 2006)], one can see the references there.

the fermion species). We would like to add that, for the sake of brevity at places, we may use $g$ instead of $g_{a\gamma\gamma}$ at some places in the rest of this paper. Therefore the additional contribution to the axion photon effective lagrangian from the new vertex would add to the existing one i.e.,eqn. [2.11].

## 3. Expression for photon axion vertex in presence of uniform background magnetic field and material medium

In order to estimate the loop induced $\gamma - a$ coupling, one can start with the Lagrangian given by Eqn. [2.9]. Defining $p' = p + k$ the effective vertex for the $\gamma - a$ coupling turns out to be,

$$i\Gamma_{\nu}(k) = g_{af}\, e\, Q_f \int \frac{d^4p}{(2\pi)^4} k^{\mu} \mathrm{Tr}\left[\gamma_{\mu}\gamma_5 iS(p)\gamma_{\nu} iS(p')\right]. \tag{3.13}$$

The effective vertex given by [3.13], is computed from the diagram given in [Fig.1]. In eqn. [3.13] $S(p)$ is the in medium fermionic propagator in external magnetic field, computed to all orders in field strength. The structure of the same can be found in [(A. K. Ganguly , 2006)]. One can easily recognize that, eqn. [3.13], has the following structure, $\Gamma_{\nu}(k) = k^{\mu}\Pi^A_{\mu\nu}(k)$. Where $\Pi^A_{\mu\nu}$, is the axial polarization tensor, comes from the axial coupling of the axions to the leptons and it's:

$$i\Pi^A_{\mu\nu}(k) = g_{af}\, e\, Q_f \int \frac{d^4p}{(2\pi)^4} \mathrm{Tr}\left[\gamma_{\mu}\gamma_5 iS(p)\gamma_{\nu} iS(p')\right]. \tag{3.14}$$

In general the axial polarization tensor, $\Pi^A_{\mu\nu}$ (some times called the VA response function), would have contributions from pure magnetic field background, as well as magnetic field plus medium, i.e., magnetized medium. The contribution from only magnetic field and the one with magnetized medium effects, are given in the following expression,

$$i\Pi^A_{\mu\nu}(k) = g_{af}\, e\, Q_f \int \frac{d^4p}{(2\pi)^4} \mathrm{Tr}\Big[\gamma_{\mu}\gamma_5 iS^V_B(p)\gamma_{\nu} iS^V_B(p') + \gamma_{\mu}\gamma_5 S^{\eta}_B(p)\gamma_{\nu} iS^V_B(p')$$
$$+ \gamma_{\mu}\gamma_5 iS^V_B(p)\gamma_{\nu} S^{\eta}_B(p')\Big]. \tag{3.15}$$

The pure magnetic field contribution to $\Pi^A_{\mu\nu}(k)$ has been estimated in [(A. K. Ganguly , 2006; D. V. Galtsov , 1972; L. L.DeRaad *et al.* , 1976; A. N. Ioannisian et al. , 1997; C. Schubert , 2000)]. The expression of the would be provided in the next section, after that the thermal part contribution to the same would be reported .

### 3.1 Magnetized vacuum contribution

The VA response function in a magnetic field $\Pi^A$ has been evaluated in [(A. K. Ganguly , 2006; D. V. Galtsov , 1972; L. L.DeRaad *et al.* , 1976; A. N. Ioannisian et al. , 1997; C. Schubert , 2000)], with varying choice of metric; we have reevaluated it according to our metric convention $g_{\mu\nu} \equiv \mathrm{diag}\,(+1, -1, -1, -1)$. The expression for the same according our convention is:

$$\Pi^{A}_{\mu\nu}(k) = \frac{ig_{af}(eQ_f)^2}{(4\pi)^2}\int_0^\infty dt \int_{-1}^{+1} dv\, e^{\phi_0}\Big\{\Big(\frac{1-v^2}{2}k_\parallel^2 - 2m_e^2\Big)\widetilde{F}_{\mu\nu} - (1-v^2)k_{\mu_\parallel}(\widetilde{F}k)_\nu + R\Big[k_{\nu_\perp}(k\widetilde{F})_\mu + k_{\mu_\perp}(k\widetilde{F})_\nu\Big]\Big\}, \quad (3.16)$$

Where, $R = \left[\frac{1 - v\sin\mathcal{Z}v\sin\mathcal{Z} - \cos\mathcal{Z}\cos\mathcal{Z}v}{\sin^2\mathcal{Z}}\right]$ and $\phi_0 = it\left[\frac{1-v^2}{4}k_\parallel^2 - m^2 - \frac{\cos v\mathcal{Z} - \cos\mathcal{Z}}{2\mathcal{Z}\sin\mathcal{Z}}k_\perp^2\right]$. In the above expression, $\widetilde{F}^{\mu\nu} = \frac{1}{2}\epsilon^{\mu\nu\rho\sigma}F_{\rho\sigma}$, and $\epsilon^{0123} = 1$ is the dual of the field-strength tensor, with $\mathcal{Z} = eQ_f\mathcal{B}t$. Therefore, following eqn. [3.13], the photon axion vertex in a purely magnetized vacuum, would be, $\Gamma^\nu(k) = k^\mu \Pi^{A_B}_{\mu\nu}(k)$ i.e.,

$$\Gamma^\nu(k) = \frac{ig_{af}(eQ_f)^2}{(4\pi)^2}\int_0^\infty dt \int_{-1}^{+1} dv\, e^{\phi_0}k^\mu\Big\{\Big(\frac{1-v^2}{2}k_\parallel^2 - 2m_e^2\Big)\widetilde{F}_{\mu\nu} - (1-v^2)k_{\mu_\parallel}(\widetilde{F}k)_\nu + R\Big[k_{\nu_\perp}(k\widetilde{F})_\mu + k_{\mu_\perp}(k\widetilde{F})_\nu\Big]\Big\}, \quad (3.17)$$

This result is not gauge invariant. However following [(A. K. Ganguly , 2006; A. N. Ioannisian et al. , 1997)], one may integrate the first term under the integral, and arrive at the expression for, the Effective Lagrangian for loop induced axion photon coupling in a magnetized vacuum, to be given by,

$$\mathcal{L}^B_{a\gamma} = aA^\nu\Gamma_\nu(k) \quad (3.18)$$

In eqn.[3.18],we define the axion field by $a$ and $(k\widetilde{F})^\nu = k_\mu\widetilde{F}^{\mu\nu}$ and $(\widetilde{F}k)^\nu = \widetilde{F}^{\nu\mu}k_\mu$. Finally the loop induced contribution to the axion photon effective Lagrangian is,

$$\mathcal{L}^{\mathcal{B}}_{a\gamma} == -\frac{1}{32\pi^2}g_{af}(eQ_f)^2\left[4 + \frac{4}{3}\left(\frac{k_\parallel^2}{m^2}\right)\right]aF_{\mu\nu}\tilde{F}^{\mu\nu}. \quad (3.19)$$

Since we are interested in $\omega < m$, so the magnitude of the factor $\left(\frac{k_\parallel}{m}\right)^2 << 1$, thus the order of magnitude estimate estimate of this contribution is of $O(1)$. However some of the factors there are momentum dependent, so it may affect the dispersion relation for photon and axion.

## 4. Contribution from the magnetized medium

Having estimated the effective axion photon vertex in a purely magnetic environment, we would focus on the contribution from the magnetized medium. As before, one can evaluate the same by using the expression for a fermion propagator in external magnetic field and medium; the result is:

$$\begin{aligned}\Pi^{A_\beta}_{\mu\nu}(k) &= (ig_{af}eQ_f)\int\frac{d^4p}{(2\pi)^4}\int_{-\infty}^{\infty}ds\,e^{\Phi(p,s)}\int_0^\infty ds' e^{\Phi(p',s')}\mathrm{Tr}\Big[ \\ &\quad [\gamma_\mu\gamma_5 G(p,s)\gamma_\nu G(p',s')]\,\eta_F(p) + [\gamma_\mu\gamma_5 G(-p',s')\gamma_\nu G(-p,s)]\,\eta_F(-p)\Big] \\ &= (ig_{af}eQ_f)\int\frac{d^4p}{(2\pi)^4}\int_{-\infty}^{\infty}ds\,e^{\Phi(p,s)}\int_0^\infty ds'\,e^{\Phi(p',s')}\mathrm{R}_{\mu\nu}(p,p',s,s') \quad (4.20)\end{aligned}$$

where $\mathrm{R}_{\mu\nu}(p,p',s,s')$ contains the trace part. $\mathrm{R}_{\mu\nu}(p,p',s,s')$ is a polynomial in powers of the external magnetic field with even and odd powers of $\mathcal{B}$, can be presented as,

$$\mathrm{R}_{\mu\nu}(p,p',s,s') = \mathrm{R}^{(E)}_{\mu\nu}(p,p',s,s') + \mathrm{R}^{(O)}_{\mu\nu}(p,p',s,s') \tag{4.21}$$

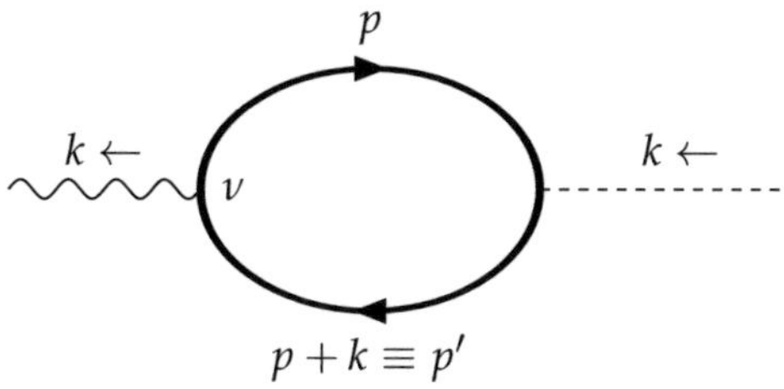


Fig. 1. One-loop diagram for the effective axion electromagnetic vertex .

We have denoted the pieces with even and odd powers in the external magnetic field strength $\mathcal{B}$ in $\mathrm{R}_{\mu\nu}$, as $\mathrm{R}^{(E)}_{\mu\nu}$ and $\mathrm{R}^{(O)}_{\mu\nu}$. In addition to being just even and odd in powers of $eQ_f\mathcal{B}$, they are also odd and even in powers of chemical potential, therefore, under charge conjugation they would transform as, $\mathcal{B}\&\mu \leftrightarrow (-\mu)\&(-\mathcal{B})$,i. e., both behave differently. More over their parity structures are also different. These properties come very useful while analyzing, the structure of axion photon coupling, using discrete symmetry arguments to justify the presence or absence of either of the two; that is the reason, why they should be treated separately. The details of this analysis can be found in [(A. K. Ganguly , 2006)].

#### 4.0.1 Vertex function: even powers in B

The expression for the $\mathrm{R}^{E}_{\mu\nu}$, (that is the term with even powers of the magnetic field), comes out to be,

$$\mathrm{R}^{(E)}_{\mu\nu} \overset{\circ}{=} 4i\eta_-(p_0)\Bigg[\varepsilon_{\mu\nu\alpha\beta}p^{\alpha}k^{\beta}(1+\tan(eQ_f\mathcal{B}s)\tan(eQ_f\mathcal{B}s')) + \varepsilon_{\mu\nu\alpha\beta_\perp}k^{\alpha}k^{\beta_\perp}$$
$$\times \tan(eQ_f\mathcal{B}s)\tan(eQ_f\mathcal{B}s')\frac{\tan(eQ_f\mathcal{B}s)-\tan(eQ_f\mathcal{B}s')}{\tan(eQ_f\mathcal{B}s)+\tan(eQ_f\mathcal{B}s')}\Bigg]. \tag{4.22}$$

Because of the presence of $\varepsilon_{\mu\nu\alpha\beta}k^{\beta}$ and $\varepsilon_{\mu\nu\alpha\beta_\perp}k^{\alpha}$, it vanishes on contraction $\mathrm{R}^{(E)}_{\mu\nu}$ with $k^{\nu}$.

The two point VA response function $\Pi^{A}(k)$, can be interpreted as a (one particle irreducible) two point vertex; with one point for the external axion line and the other one (Lorentz indexed) for the external photon line. But since the evaluations are done in presence of external magnetic field $\mathcal{B}$ they correspond to soft external photon line insertions. That is their four momenta $k^{\alpha} \to 0$ . If each soft external photon line contributes either +1 or -1 to the total spin ( angular momentum ) of the effective vertex, then, for an even order term in external field strength $\mathcal{B}$ the total spin of this piece would be a coherent sum of all the contributions from all the odd number of soft photon lines$\mathcal{B}$. Now recall that in order arrive at the the expression for the effective interaction Lagrangian for $\gamma - a$ from $\Pi^{A}_{\mu\nu}(x)$–we need to multiply the same (with some sort of naivete) by $a(x)F^{\mu\nu}(x)$ . Therefore, it is worth noting that, if we multiply $\Pi^{A\text{ ( Even B)}}_{\mu\nu}(x)$ with $a(x)F^{\mu\nu}(x)$, the number of photon lines become odd and number of spin zero pseudoscalar is also odd. Since the effective Lagrangian can be related to the generating functional of the vertex for transition of photons to axion, then for this case it would mean, odd number of photons are going to produce a spin zero pseudoscalar. That is odd number of spin one photons would combine to produce a spin zero axion— *which is impossible,* hence such a term better not exist. Interestingly enough, *that is what* we get to see here.

#### 4.0.2 Vertex function: odd powers in B

The nonzero contribution to the vertex function would be coming from $\mathrm{R}^{O}_{\mu\nu}$. More precisely, from the following term,

$$ik^{\mu}\mathcal{R}^{(O)}_{\mu\nu} = 8m^2\eta_{+}(p)\left[k^{\mu}\varepsilon_{\mu\nu 12}(\tan(eQ_f\mathcal{B}s) + \tan(eQ_f\mathcal{B}s'))\right], \tag{4.23}$$

Placing all the factors and integral signs, the vertex function $\Gamma_{\nu}(k)$ can be written as,

$$\begin{aligned}\Gamma_{\nu}(k) = (g_{af}eQ_f)\left(8m^2k^{\mu}\varepsilon_{\mu\nu 12}\right)\int\frac{d^4p}{(2\pi)^4}\eta_{+}(p)\int_{-\infty}^{\infty}ds\int_{0}^{\infty}ds' e^{\Phi(p,s)+\Phi(p',s')} \\ \times\left[\tan(eQ_f\mathcal{B}s) + \tan(eQ_f\mathcal{B}s')\right]\end{aligned} \tag{4.24}$$

Upon performing the gaussian integrals for the perpendicular momentum components, there after taking limit $|k| \to 0$ and assuming photon energy $\omega < m_f$ one arrives at,

$$\Gamma_{\nu}(k) == -16(g_{af}(eQ_f)^2)\left(\frac{k^{\mu}\tilde{\mathrm{F}}_{\mu\nu}}{16\pi^2}\right)\Lambda(k^2_{\parallel}, k\cdot u, \beta, \mu). \tag{4.25}$$

All the informations about the medium, are contained in $\Lambda(k^2_{\parallel}, k_{\parallel}\cdot u, \beta, \mu)$ and it is given by.

$$\Lambda(k^2_{\parallel}, k\cdot u, \beta, \mu) = \int d^2p_{\parallel}\left[n_F(|p_0|, \mu) + n_F(|p_0|, -\mu)\right]\left(\frac{m^2\delta(p^2_{\parallel} - m^2)}{(k^2_{\parallel} + 2(p\cdot k)_{\parallel})}\right) \tag{4.26}$$

In the expression above the temperature of the medium ( $\beta = 1/T$), number density of the fermions (which in turn is related to $\mu$), mass of the particles in the loop (m), energy and longitudinal momentum of the photon ( i.e. $k_{||}$). The statistical factor has already been evaluated in [(A. K. Ganguly , 2006)], in various limits. So instead of providing the same we state the result obtained in the limits $m \ll \mu$, and limit$\mathrm{T} \to 0$. The value of the same in this limit is

$$\mathrm{Lt}_{T\to 0}\Lambda \simeq -\frac{1}{2}\frac{|\frac{\mu}{m}|}{\sqrt{\left(1 + [\frac{\mu}{m}]^2\right)}} \tag{4.27}$$

In the limit $\mu \gg m$, the right hand side of Eqn. [4.27] $\sim \frac{1}{2}$ and when $\mu \sim m$, it would turn out to be $\sim \frac{1}{2\sqrt{2}}$

In the light of these estimates, it is possible to write down the axion photon mixing Lagrangian, for low frequency photons in an external magnetic field, in the following way:

$$\mathcal{L}^{Total}_{a\gamma} = \mathcal{L}^{vac}_{a\gamma} + \mathcal{L}^{\mathcal{B}}_{a\gamma} + \mathcal{L}^{\mathcal{B},\mu,\beta}_{a\gamma}. \tag{4.28}$$

Where each of the terms are given by,

$$\begin{aligned}
\mathcal{L}^{vac}_{a\gamma} &= -g_{a\gamma\gamma}\frac{e^2}{32\pi^2}a\mathrm{F}\tilde{\mathrm{F}}, \\
\mathcal{L}^{\mathcal{B}}_{a\gamma} &= \frac{-1}{32\pi^2}\left[4+\frac{4}{3}\left(\frac{k_{\parallel}}{m}\right)^2\right]\sum_f g_{af}(eQ_f)^2 a\mathrm{F}\tilde{\mathrm{F}}. \\
\mathcal{L}^{\mathcal{B},\mu,\beta}_{\gamma a} &= \frac{32}{32\pi^2}\cdot\left(\frac{k_{\parallel}}{\omega}\right)^2(\Lambda)\sum_f g_{af}(eQ_f)^2 a\mathrm{F}\tilde{\mathrm{F}}.
\end{aligned} \tag{4.29}$$

Therefore, in the limit of $|k_{\perp}| \to 0$ and $\omega << m_f$, one can write the total axion photon effective Lagrangian using eqn. [4.29], in the following form.

$$\mathcal{L}^{Total}_{a\gamma} = -\left[g_{a\gamma\gamma} + \left(4+\frac{4}{3}\left(\frac{k_{\parallel}}{m}\right)^2\right)\sum_f g_{af}(Q_f)^2 - 32\left(\frac{k_{\parallel}}{\omega}\right)^2\Lambda\sum_f g_{af}(Q_f)^2\right]\frac{e^2}{32\pi^2}a\mathrm{F}\tilde{\mathrm{F}}. \tag{4.30}$$

We would like to point out that, the in medium corrections doesn't alter the tensorial structure of the same. It remains intact. However the parameter $M$ , doesn't remain so. Apart from numerical factors it also starts depending on the kinematic factors. It is worth noting that, all the terms generated by loop induced corrections do respect **CPT**. Additionally, as we have analyzed already the total spin angular momentum is also conserved. The tree level photon axion interaction term in the Lagrangian as found in the literature is of the following form,

$$\frac{1}{M}a\mathrm{F}^{\mu\nu}.\tilde{F}^{ext}_{\mu\nu}, \tag{4.31}$$

The bounds on various axion parameters are obtained by using this Lagrangian. As we have seen the medium and other corrections can affect the magnitude of $M$ . Since $M$ is related to the symmetry breaking scale, a change in the estimates of $M$ would have reflection on the symmetry breaking scale and other axion parameters. This is the primary motivation for our dwelling on this part of the problem before moving into aspects of axion electrodynamics, that affects photon polarization.

## 5. Axion photon mixing

Now that we are equipped with the necessary details of axion interactions with other particles, we can write down the relevant part of the Lagrangian that describes the Axion photon interaction. The tree level Lagrangian that describes the axion photon dynamics is given by,

$$\mathcal{L} = -\frac{1}{4}\mathrm{F}^{\mu\nu}\mathrm{F}_{\mu\nu} + \frac{1}{4M}\mathrm{F}^{\mu\nu}\tilde{\mathrm{F}}_{\mu\nu} + \frac{1}{2}\left(\partial_\mu a\partial^\mu a - m_a^2a^2\right), \tag{5.32}$$

here $m_a$ , is the axion mass and other quantities have their usual meaning. This effective Lagrangian shows the effect of mixing of a spin zero pseudo-scalar with two photons. If one of the dynamical photon field in eqn. [5.32] is replaced by an external magnetic field, one would recover the Lagrangian given by eqn.[4.31]. This mixing part can give rise to various interesting observable effects; however in this section we would consider, the change in the state of polarization of a plane polarized light beam, propagating in an external magnetic field, due to axion photon mixing. In order to perform that analysis, we start with the equation of motion for the photons and the axions, in an external magnetic field $\mathcal{B}$ , that follows from the interaction part of the Lagrangian in eqn. [5.32], as we replace one of the dynamical photon field by external magnetic field field.

This system that we are going to study involve the dynamics of three field Degrees Of Freedom (DOF). As we all know, that the massless spin one gauge fields in vacuum have just two degrees of freedom; so we have those two DOF and the last one is for the spin zero pseudoscalar Boson. In this simple illustrative analysis, we would ignore the transverse component of the momentum $k_\perp$. With this simplification in mind we have three equations of motion, one each for: $A_\perp(z)$, $A_{||}(z)$ and $a(z)$–i.e., the three dynamical fields. Where $A_\perp(z)$ , the photon/gauge field with polarization vector directed along the perpendicular direction to the magnetic field, $A_{||}(z)$ the remaining component of the photon/gauge field having polarization vector lying along the magnetic field and $''a(z)''$ the pseudoscalar Axion field. These three equations can be written in a compcat form e g.,

$$\left[(\omega^2+\partial_z^2)\ \mathbf{I}+\mathcal{M}\right]\begin{pmatrix}A_\perp\\ A_{||}\\ a\end{pmatrix}=0. \tag{5.33}$$

where **I** is a $3\times 3$ identity matrix and **M** is the short hand notation for the following matrix.

$$\mathcal{M}=\begin{pmatrix}0&0&0\\0&0&ig\mathcal{B}\omega\\0&-ig\mathcal{B}\omega&-m_a^2\end{pmatrix}, \tag{5.34}$$

usually termed as axion photon mixing matrix or simply the mixing matrix. As can be seen from eqn.[5.33], the transverse gauge degree of freedom gets decoupled from the rest, and the other two i.e., the longitudinal gauge degrees of freedom and pseudoscalar degree of freedom are coupled with each other. It is because of this particular way of evolution of the transverse and the parallel components of the gauge field, even magnetized vacuum would show dichoric effect.

In the off diagonal element of the matrix [5.34] given by, $\pm ig\mathcal{B}\omega$, $\mathcal{B}=\mathcal{B}^E\sin\ (\hat{\alpha})$, is the transverse part of the external magnetic field $\mathcal{B}^E$ and $\hat{\alpha}$ is the angle between the wave vector $\vec{k}$ and the external magnetic field $\mathcal{B}^E$ and lastly in a short hand notation, $g=\frac{1}{\mathrm{M}}$. The nondiagonal part of the 3x3 matrix, in eqn. [5.34] can be written as,

$$M_{2\times 2}=\begin{pmatrix}0&ig\mathcal{B}\omega\\-ig\mathcal{B}\omega&-m_a^2\end{pmatrix}. \tag{5.35}$$

One can solve for the eigen values of the eqn. [5.35], from the determinantal equation,

$$\begin{vmatrix} \mathrm{M}_j & -ig\mathcal{B}\omega \\ ig\mathcal{B}\omega & m_a^2 + \mathrm{M}_j \end{vmatrix} = 0. \tag{5.36}$$

In eqn. [5.36] $j$ can take either of the two values $+$ or $-$, and the roots are as follows:

$$M_{\pm} = -\frac{m_a^2}{2} \pm \sqrt{\left[\left(\frac{m_a^2}{2}\right)^2 + (g\mathcal{B}\omega)^2\right]}. \tag{5.37}$$

## 6. Equation of motion

The equations of motion for the photon field with polarization vector in the perpendicular direction to the external magnetic filed is,

$$\left[(\omega^2 + \partial_z^2)\right]\left(A_{\perp}\right) = 0\,. \tag{6.38}$$

The remaining single physical degree freedom for the photon, with polarization along the external magnetic field, gets coupled with the axion; and the equation of motion turns out to be,

$$\left[(\omega^2 + \partial_z^2)\ \mathbf{I} + M_{2\times 2}\right]\begin{pmatrix} A_{\|} \\ a \end{pmatrix} = 0. \tag{6.39}$$

It is possible to diagonalize eqn.[6.39] by a similarity transformation. We would denote the diagonalizing matrix by $O$, given by,

$$O = \begin{pmatrix} \cos\theta & -\sin\theta \\ \sin\theta & \cos\theta \end{pmatrix} \equiv \begin{pmatrix} c & -s \\ s & c \end{pmatrix}, \tag{6.40}$$

in short. The diagonal matrix can further be written as,

$$M_D = O^T M_{2\times 2} O = \begin{pmatrix} c & s \\ -s & c \end{pmatrix}\begin{pmatrix} M_{11} & M_{12} \\ M_{21} & M_{22} \end{pmatrix}\begin{pmatrix} c & -s \\ s & c \end{pmatrix}, \tag{6.41}$$

with the following forms for the elements of the matrix $\mathrm{M}_{2\times 2}$, given by: $M_{11} = 0$, $M_{12} = ig\mathcal{B}\omega$, $M_{21} = -ig\mathcal{B}\omega$ and lastly $M_{22} = -m_a^2$. The value of the parameter $\theta$ is fixed from the equality,

$$M_D = \begin{pmatrix} c & s \\ -s & c \end{pmatrix}\begin{pmatrix} M_{11} & M_{12} \\ M_{21} & M_{22} \end{pmatrix}\begin{pmatrix} c & -s \\ s & c \end{pmatrix} = \begin{pmatrix} M_+ & 0 \\ 0 & M_- \end{pmatrix}, \tag{6.42}$$

leading to,

$$\begin{pmatrix} c^2M_{11} + s^2M_{22} + 2csM_{12} & M_{12}(c^2 - s^2) + cs(M_{22} - M_{11}) \\ M_{12}(c^2 - s^2) + cs(M_{22} - M_{11}) & s^2M_{11} + c^2M_{22} - 2csM_{12} \end{pmatrix} = \begin{pmatrix} M_+ & 0 \\ 0 & M_- \end{pmatrix}, \tag{6.43}$$

Now equating the components of the matrix equation [6.43], one arrives at:

$$\tan(2\theta) = \frac{2M_{12}}{M_{11} - M_{22}} = \frac{2ig\mathcal{B}\omega}{m_a^2}. \tag{6.44}$$

Therefore upon using this similarity transformation, the coupled Axion photon differential equation can further be brought to the following form,

$$\left[ (\omega^2 + \partial_z^2)\ \mathbf{I} + \ M_D \right] \begin{pmatrix} \bar{A}_{\|} \\ \bar{a} \end{pmatrix} = 0. \tag{6.45}$$

## 7. Dispersion relations

Defining the wave vectors in terms of $k_i$'s, as:

$$\begin{aligned} k_{\perp} &= \omega \\ k_{+} &= \sqrt{\omega^2 + M_{+}} \\ k_{-} &= -\sqrt{\omega^2 + M_{+}} \end{aligned} \tag{7.46}$$

and

$$\begin{aligned} k'_{+} &= \sqrt{\omega^2 + M_{-}} \\ k'_{-} &= -\sqrt{\omega^2 + M_{-}} \end{aligned} \tag{7.47}$$

## 8. Solutions

The solutions for the gauge field and the axion field, given by [6.45] as well as the solution for eqn. for $A_{\perp}$ in $k$ space can be written as,

$$\bar{A}_{||}(z) = \bar{A}_{||_+}(0)e^{ik_+z} + \bar{A}_{||_-}(0)e^{-ik_-z} \tag{8.48}$$

$$\bar{a}(z) = \bar{a}_{+}(0)\ e^{ik'_+z} + \bar{a}_{-}(0)\ e^{-ik'_-z} \tag{8.49}$$

$$A_{\perp}(z) = A_{\perp_+}(0)e^{ik_{\perp}z} + A_{\perp_-}(0)e^{-ik_{\perp}z} \tag{8.50}$$

## 9. Correlation functions

The solutions for propagation along the +ve z axis, is given by,

$$\bar{A}_{||}(z) = \bar{A}_{||_+}(0)e^{ik_+z} \tag{9.51}$$

$$\bar{a}(z) = \bar{a}_{+}(0)\ e^{ik'_+z} \tag{9.52}$$

that can further be written in the following form,

$$\begin{pmatrix} \bar{A}_{||}(z) \\ \bar{a}(z) \end{pmatrix} = \begin{pmatrix} e^{ik_+z} & 0 \\ 0 & e^{ik'_+z} \end{pmatrix} \begin{pmatrix} \bar{A}_{||}(0) \\ \bar{a}(0) \end{pmatrix}. \tag{9.53}$$

Since,

$$\begin{pmatrix} \bar{A}_{||}(z/0) \\ \bar{a}(z/0) \end{pmatrix} = O^T \begin{pmatrix} A_{||}(z/0) \\ a(z/0) \end{pmatrix}. \tag{9.54}$$

it follows from there that,

$$\begin{pmatrix} A_{||}(z) \\ a(z) \end{pmatrix} = O \begin{pmatrix} e^{ik_+ z} & 0 \\ 0 & e^{ik'_+ z} \end{pmatrix} O^T \begin{pmatrix} A_{||}(0) \\ a(0) \end{pmatrix}. \tag{9.55}$$

Using eqn.[9.55] we arrive at the relation,

$$A_{||}(z) = \left[e^{ik_+ z}\cos^2\theta + e^{ik'_+ z}\sin^2\theta\right] A_{||}(0) + \left[e^{ik_+ z} - e^{ik'_+ z}\right] \cos\theta \sin\theta \, a(0) \tag{9.56}$$

$$a(z) = \left[e^{ik_+ z} - e^{ik'_+ z}\right] \cos\theta \sin\theta \, A_{||}(0) + \left[e^{ik_+ z}\sin^2\theta + e^{ik'_+ z}\cos^2\theta\right] a(0) \tag{9.57}$$

If we assume the axion field to be zero, to begin with, i.e., $a(0) = 0$, then the solution for the gauge fields take the follwing form,

$$\begin{aligned} A_{||}(z) &= \left[e^{ik_+ z}\cos^2\theta + e^{ik'_+ z}\sin^2\theta\right] A_{||}(0) \\ A_\perp(z) &= e^{ik_\perp z} A_\perp(0). \end{aligned} \tag{9.58}$$

Now we can compute various correlation functions with the photon field. The correlation functions of parallel and perpendicular components of the photon field take the following form:

$$\begin{aligned} &< A^*_{||}(z) A_{||}(z) >= \left[\cos^4\theta + \sin^4\theta + 2\sin^2\theta\,\cos^2\theta\,\cos\left[(k_+ - k'_+)\,z\right]\right] < A^*_{||}(0) A_{||}(0) > \\ &< A^*_{||}(z) A_\perp(z) >= \left[\cos^2\theta e^{i(k_\perp - k_+)z} + \sin^2\theta\, e^{i(k_\perp - k'_+)z}\right] < A^*_{||}(0) A_\perp(0) > \\ &< A^*_\perp(z) A_\perp(z) >=< A^*_\perp(0) A_\perp(0) > \end{aligned} \tag{9.59}$$

## 10. Digression on stokes parameters

Various optical parameters like polarization, ellipticity and degree of polarization of a given light beam can be found from the the coherency matrix constructed from various correlation functions given above. The coherency matrix, for a system with two degree of freedom is defined as an ensemble average of direct product of two vectors:

$$\rho(z) = \langle \begin{pmatrix} A_{||}(z) \\ A_\perp(z) \end{pmatrix} \otimes \begin{pmatrix} A_{||}(z) & A_\perp(z) \end{pmatrix}^* \rangle = \begin{pmatrix} \langle A_{||}(z) A^*_{||}(z) \rangle & \langle A_{||}(z) A^*_\perp(z) \rangle \\ \langle A^*_{||}(z) A_\perp(z) \rangle & \langle A_\perp(z) A^*_\perp(z) \rangle \end{pmatrix} \tag{10.60}$$

The important thing to note here is that, under any anticlock-wise rotation $\alpha$ about an axis perpendicular the $||$ and $\perp$ components, would convert:

$$\rho(z) \to \rho'(z) = \langle \mathcal{R}(\alpha) \begin{pmatrix} A_{||}(z) \\ A_\perp(z) \end{pmatrix} \otimes \begin{pmatrix} A_{||}(z) & A_\perp(z) \end{pmatrix}^* \mathcal{R}^{-1}(\alpha) \rangle \tag{10.61}$$

where $\mathcal{R}(\alpha)$ is the rotation matrix. Now from the relations between the components of coherency matrix and the stokes parameters:

$$\begin{aligned}
&\mathrm{I}=< A_{||}^{*}(z)A_{||}(z) > + < A_{\perp}^{*}(z)A_{\perp}(z) >,\\
&\mathrm{Q}=< A_{||}^{*}(z)A_{||}(z) > - < A_{\perp}^{*}(z)A_{\perp}(z) >,\\
&\mathrm{U}=2\mathrm{Re} < \mathrm{A}_{||}^{*}(\mathrm{z})\mathrm{A}_{\perp}(\mathrm{z}) >,\\
&\mathrm{V}=2\,\mathrm{Im} < \mathrm{A}_{||}^{*}(\mathrm{z})\mathrm{A}_{\perp}(\mathrm{z}) > .
\end{aligned} \tag{10.62}$$

It is easy to establish that,

$$\rho(z) = \frac{1}{2}\begin{pmatrix} \mathrm{I}(z) + \mathrm{Q}(z) & \mathrm{U}(z) - i\mathrm{V}(z) \\ \mathrm{U}(z) + i\mathrm{V}(z) & \mathrm{I}(z) - \mathrm{Q}(z) \end{pmatrix} \tag{10.63}$$

Therefore, under an anticlock wise rotation by an angle $\alpha$, about an axis perpendicular to the plane containing $A_{||}(z)$ and $A_{\perp}(z)$, the density matrix transforms as: $\rho(z) \rightarrow \rho'(z)$; the same in the rotated frame would be given by,

$$\rho'(z) = \frac{1}{2}\mathcal{R}(\alpha)\begin{pmatrix} \mathrm{I}(z) + \mathrm{Q}(z) & \mathrm{U}(z) - i\mathrm{V}(z) \\ \mathrm{U}(z) + i\mathrm{V}(z) & \mathrm{I}(z) - \mathrm{Q}(z) \end{pmatrix}\mathcal{R}^{-1}(\alpha)\,. \tag{10.64}$$

For a rotation by an angle $\alpha$–in the anticlock direction– about an axis perpendicular to $A_{||}$ and $A_{\perp}$ plane, the rotation matrix $\mathcal{R}(\alpha)$ is,

$$\mathcal{R}(\alpha) = \begin{pmatrix} \cos\alpha & \sin\alpha \\ -\sin\alpha & \cos\alpha \end{pmatrix}. \tag{10.65}$$

From the relations above, its easy to convince oneself that, in the rotated frame of reference the two stokes parameters, Q and U get related to the same in the unrotated frame, by the following relation.

$$\begin{pmatrix} \mathrm{Q}'(z) \\ \mathrm{U}'(z) \end{pmatrix} = \begin{pmatrix} \cos 2\alpha & \sin 2\alpha \\ -\sin 2\alpha & \cos 2\alpha \end{pmatrix}\begin{pmatrix} \mathrm{Q}(z) \\ \mathrm{U}(z) \end{pmatrix} \tag{10.66}$$

The other two parameters, i.e., I and V remain unaltered. It is for this reason that some times I and V are termed invariants under rotation.

For a little digression, we would like to point out that, in a particular frame, the Stokes parameters are expressed in terms of two angular variables $\chi$ and $\psi$ usually called the ellipticity parameter and polarization angle, defined as,

$$\begin{aligned}
\mathrm{I} &= \mathrm{I}_p\\
\mathrm{Q} &= \mathrm{I}_p\cos 2\psi \cos 2\chi\\
\mathrm{U} &= \mathrm{I}_p\sin 2\psi \cos 2\chi\\
\mathrm{V} &= \mathrm{I}_p\sin 2\chi.
\end{aligned} \tag{10.67}$$

The ellipticity angle, $\chi$, following [10.67], can be shown to be equal to,

$$\tan 2\chi = \frac{V}{\sqrt{Q^2 + U^2}}, \tag{10.68}$$

and the polarization angle can be shown to be equal to.

$$\tan 2\psi = \frac{U}{Q} \tag{10.69}$$

From the relations given above, it is easy to see that, under the frame rotation,

$$\mathcal{R}(\alpha) = \begin{pmatrix} \cos 2\alpha & \sin 2\alpha \\ -\sin 2\alpha & \cos 2\alpha \end{pmatrix} \tag{10.70}$$

the Tangent of $\chi$, i.e., $\tan\chi$ remains invariant, however the tangent of the polarization angle gets additional increment by twice the rotation angle, i.e.,

$$\begin{aligned} \tan(2\chi) &\rightarrow \tan(2\chi) \\ \tan(2\psi) &\rightarrow \tan(2\alpha + 2\psi). \end{aligned} \tag{10.71}$$

It is worth noting that the two angles are not quite independent of each other, in fact they are ralated to each other. Finally we end the discussion of use of stokes parameters by noting that, the degree of polarization is usually expressed by,

$$p = \frac{\sqrt{Q^2 + U^2 + V^2}}{I_{P_T}} \tag{10.72}$$

where $I_{P_T}$ is the total intensity of the light beam.

## 11. Evaluation of ellipticity ($\chi$) and polarization ($\psi$) angles

Now we would proceed further from the formula given in the previous sections, to evaluate the ellipticity and polarization angles for a beam of plane polarized light propagating in the z direction. Since we are interested in finding out the effect of axion photon mixing, we need the expressions for the Stokes parameters with the Axion photon mixing effect and with that we would evaluate the ellipticity angle $\chi$ and polaraization angle $\psi$ at a distance $z$ from the source. Using the expressions for the correlators (i.e., eqns. [9.59] ) , one can evaluate the stokes parameters and they turn out to be

$$\begin{aligned}
I &= \left[\cos^4\theta + \sin^4\theta + 2\sin^2\theta\cos^2\theta\cos\left[\left(k_+ - k'_+\right)z\right]\right] < A^*_{||}(0)A_{||}(0) > + < A^*_{\perp}(0)A_{\perp}(0) > \\
Q &= \left[\cos^4\theta + \sin^4\theta + 2\sin^2\theta\cos^2\theta\cos\left[\left(k_+ - k'_+\right)z\right]\right] < A^*_{||}(0)A_{||}(0) > - < A^*_{\perp}(0)A_{\perp}(0) > \\
U &= 2\left(\left[\cos^2\theta\cos\left[\left(k_\perp - k_+\right)z\right]\right] + \sin^2\theta\cos\left[\left(k_\perp - k'_+\right)z\right]\right) < A^*_{||}(0)A_{\perp}(0) > \\
V &= 2\left(\left[\cos^2\theta\sin\left[\left(k_\perp - k_+\right)z\right]\right] + \sin^2\theta\sin\left[\left(k_\perp - k'_+\right)z\right]\right) < A^*_{||}(0)A_{\perp}(0) >
\end{aligned} \tag{11.73}$$

Till this point, the expressions, we obtain are very general i. e., no approximations were made. However for predicting or explaining the experimental outcome one would have to choose some initial conditions and make some approximations to evaluate the physical quantities of interest. In that spirit, in this analysis we would take the initial beam of light to be plane polarized, with the plane of polarization making an angle $\frac{\pi}{4}$ with the external magnetic field. And their amplitude would be assumed to be unity; therefore under this approximation $A_{||}(0) = A_{\perp}(0) = \frac{1}{\sqrt{2}}$.

It is important to note that, for axion detection through polarization measurements or, astrophysical observations, the parameter $\theta << 1$. Also we can define another dimension full parameter, $\delta = \frac{g}{m_a^2}$. With the current experimental bounds for Axion mass and coupling constant $\delta << 1$. So we can safely take $\cos\theta \sim 1$ and $\sin\theta \sim \theta$. Now going back to eqns., (7.46) and (7.47) one can see that the dispersion relations for the wave vectors are given by,

$$\begin{aligned} k_{\perp} &\simeq \omega, \\ k_{+} &\simeq \omega + \frac{(g\mathcal{B}\omega)^2}{2m_a^2\omega}, \\ k'_{+} &\simeq \omega - \frac{m_a^2}{2\omega} - \frac{(g\mathcal{B}\omega)^2}{2m_a^2\omega} \\ \theta &= \frac{g\mathcal{B}\omega}{m_a^2} \end{aligned} \tag{11.74}$$

Since the ratio $\frac{g}{m_a^2} = \delta << 1$, we can always neglect their higher order contributions in any expansion involving $\delta$. Therefore making the same, $Q$ can be shown to be close to zero and the Stokes parameter U turns out to be:

$$\mathrm{U} = 1 + O(\delta^n) \text{ when } n \geq 1 \,.. \tag{11.75}$$

Before proceeding further, we note the following relations,

$$\begin{aligned} k_{+} - k_{\perp} &= \frac{m_a^2\theta^2}{2\omega} \\ k'_{+} - k_{\perp} &= -\frac{m_a^2}{2\omega}, \\ k_{+} - k'_{+} &\simeq \frac{m_a^2}{2\omega} \,. \end{aligned} \tag{11.76}$$

they would be useful to find out the other Stokes parameter V. In terms of these, V comes out to be,

$$\mathrm{V} = \sin(-\frac{m_a^2\theta^2 z}{2\omega}) + \theta^2\sin(m_a^2 z/2\omega) \tag{11.77}$$

If we retain terms of order $\theta^2$ only, in eqn. [11.77], then, we find, $\mathrm{V} = \frac{1}{48}\frac{\theta^2 m_a^6 z^3}{\omega^3}$, where an overall sign has been ignored. Finally substituting the values of $\theta$ and other quantities, the ellipticity angle $\chi$ is turns out to be

$$\chi = \frac{1}{96\omega}\left(\frac{(\mathcal{B}m_a^2)}{M}\right)^2 z^3. \tag{11.78}$$

The expression of the ellipticity angle $\chi$ as given by eqn. [11.78], found to be consistent with the same in (R. Cameron et al. , 1993). It should however be noted that, for interferometer based experiments, if the path length between the mirrors is given by $l$, and there are n reflections that take place between the mirrors then $\chi(nl) = n\chi(l)$, i.e. the coherent addition of ellipticity per-pass. The reason is the following: every time the beam falls on the mirror the photons get reflected, the axions are lost, they don't get reflected from the mirror.

Having evaluated the ellipticity parameter, we would move on to calculate the polaraization angle from the expression

$$\tan(2\psi) = \frac{\mathrm{U}}{\mathrm{Q}}.$$

However there is little subtlety involved in this estimation; recall that the beam is initially polarized at an angle $45^0$ with the external magnetic field. So to find out the final polarization after it has traversed a length z, we need to rotate our coordinate system by the same angle and evaluate the cumulative change in the polarization angle. We have already noted in the previous section, the effect of such a rotation on the stokes parameters and hence on the polarization angle; so following the same procedure, we evaluate the angle $\Psi$ from the following relation,

$$\tan(2\psi + \frac{\pi}{2}) = \frac{\mathrm{U}}{\mathrm{Q}}. \tag{11.79}$$

We have already noted (eq. [11.75]) that for the magnitudes of the parameters of interest, the stokes parameter U $\sim 1$; and that makes the angle $2\psi$ inversely proportional to Q, where the proportionality constant turns out to be unity. Therefore we need to evaluate just Q, using the approximations as stated before. Recalling the fact that, the mixing angle $\theta$ is much less than one, we can expand all the $\theta$ dependent terms in the expression for Q, and retain terms up to order $\theta^2$. Once this is done, we arrive at:

$$\mathrm{Q} = -2\theta^2\left(\sin^2\left(\frac{(k_+ - k'_+)z}{2}\right)\right), \tag{11.80}$$

Now one can substitute the necessary relations given in eqns. [11.77] in eqn. [11.80] to arrive at the expression for $\psi$. Once substituted the polarization angle turns out to be.

$$\psi = \frac{(\mathcal{B}^E z)^2}{16M^2\omega}. \tag{11.81}$$

We would like to point out that, the angle of polarization as given by [11.81] also happens to be consistent with the same given in reference [(R. Cameron et al. , 1993)] where the authors had evaluated the same using a different method. In the light of this, we conclude this section by noting that, all the polarization dependent observables related to optical activity can be obtained independently by various methods, for the parameter ranges of interest or instrument sensitivity, the results obtained using stokes parameters turns out to be consistent with the alternative ones.

## 12. Axion electrodynamics in a magnetized media

In the earlier section we have detailed the procedure of getting axion photon modified equation of presence of tree level axion photon interaction Lagrangian. And this equation of motion would be valid in vacuum, but in nature most of the physical processes take place in the presence of a medium, ideal vacuum is hardly available. Therefore to study the axion photon system and their evolution one needs to take the effect of magnetized vacuum into account. This could be done by taking an effective Lagrangian, that incorporates the magnetized matter effects. This Lagrangian is provided in [(A. K. Ganguly P.K. Jain and S. Mandal , 2009)].

In momentum space this effective Lagrangian is given by:,

$$\mathcal{L} = \frac{1}{2}\left[-A_\mu k^2 \tilde{g}^{\mu\nu} A_\nu + A_\mu \tilde{\Pi}^{\mu\nu} A_\nu + i\frac{\tilde{\mathcal{F}}^{\mu\nu} k_\mu A_\nu a}{M_a} - a(k^2 - m_a^2)a\right]. \tag{12.82}$$

The notations in eqn. [12.82] are the following, $\tilde{g}^{\mu\nu} = \left(g^{\mu\nu} - \frac{k^\mu k^\nu}{k^2}\right)$, $\tilde{\mathcal{F}}^{\mu\nu}$ is the field strength tensor of the external field, $\frac{1}{M_a} \simeq \frac{1}{M}$ the axion photon coupling constant, $\tilde{\Pi}^{\mu\nu}$ is polarization tensor including Faraday contribution and is given by,

$$\tilde{\Pi}^{\mu\nu}(k) = \Pi_T(k)R^{\mu\nu} + \Pi_L(k)Q^{\mu\nu}(k) + \Pi_p(k)P^{\mu\nu}. \tag{12.83}$$

Usually in the thermal field theory notations, the cyclotron frequency is given by, $\omega_B = \frac{eB}{m}$ and plasma frequency (in terms of electron density $n_e$ and temperature $T$) in written as, $\omega_p = \sqrt{\frac{4\pi\alpha n_e}{m}\left(1 - \frac{5T}{2m}\right)}$. In terms of these expressions, the longitudinal form factor $\Pi_L$ , transverse form factor $\Pi_T$ and Faraday form factor $\Pi_p$ along with their projection operators $Q^{\mu\nu}$, $R^{\mu\nu}$ and $P^{\mu\nu}$ are given by,

$$\Pi_L(k) = k^2\omega_p^2\left(\frac{1}{\omega^2} + 3\frac{|\vec{k}|^2}{\omega^4}\frac{T}{m}\right),\quad \Pi^p(k) = \frac{\omega\omega_B\omega_p^2}{\omega^2 - \omega_B^2} \text{ and } \Pi_T = \omega_p^2\left(1 + \frac{|\vec{k}|^2}{\omega^2}\frac{T}{m}\right)$$

$$\text{where}\quad \begin{cases} Q_{\mu\nu} = \frac{\tilde{u}_\mu \tilde{u}_\nu}{\tilde{u}^2} \\ R_{\mu\nu} = \tilde{g}_{\mu\nu} - Q_{\mu\nu}, \\ P_{\mu\nu} = i\epsilon_{\mu_\perp \nu\alpha\beta}\frac{k^\alpha}{|K|}u^\beta. \end{cases}$$

The equations of motion for Gauge pseudoscala fields that follows from the Lagrangian (12.82) are the following:

$$\left(-k^2\tilde{g}_{\alpha\nu} + \tilde{\Pi}_{\alpha\nu}(k)\right)A^\nu(k) = -i\frac{k^\mu \tilde{\mathcal{F}}_{\mu\alpha} a}{2M_a} \tag{12.84}$$

$$\left(k^2 - m^2\right)a = i\frac{b_\mu^{(2)} A^\mu(k)}{2M_a}. \tag{12.85}$$

For the problem in hand we have two vectors and one tensor at our disposal, frame velocity of the medium $u^\mu$, 4 momentum of the photon $k^\mu$ and external magnetic field strength tensor $\mathcal{F}^{\mu\nu}$. To describe the dynamics of the 4 component gauge field, we need to expand them in an orthonormal basis. One can construct the basis in terms of the following 4-vectors,:

$$b^{(1)\nu} = k_\mu \mathcal{F}^{\mu\nu}, \quad b^{(2)\nu} = k_\mu \tilde{\mathcal{F}}^{\mu\nu}, \quad I^\nu = \left( b^{(2)\nu} - \frac{(\tilde{u}^\mu b^{(2)}_\mu)}{\tilde{u}^2} \tilde{u}^\nu \right), \quad \text{and } k^\mu. \tag{12.86}$$

In eqn. [12.86] we have made use of the additional vector, $\tilde{u}^\nu = \tilde{g}^{\nu\mu} u_\mu$ $(u^\mu = (1,0,0,0))$.

$$\begin{aligned} \mathrm{N}_1 &= \frac{1}{\sqrt{-b^{(1)}_\mu b^{(1)\mu}}} = \frac{1}{B_z K_\perp} \\ \mathrm{N}_2 &= \frac{1}{\sqrt{-I_\mu I^\mu}} = \frac{|\vec{K}|}{\omega K_\perp B_z} \\ \mathrm{N}_L &= \frac{1}{\sqrt{-\tilde{u}_\mu \tilde{u}^\mu}} = \frac{K}{|\vec{K}|}, \end{aligned} \tag{12.87}$$

The negative sign under the square roots are taken to make the vectors real. The Gauge field or photon field now can be expanded in this new basis,

$$A_\alpha(k) = A_1(k)\mathrm{N}_1 b^{(1)}_\alpha + A_2(k)\mathrm{N}_2 I_\alpha + A_L(k)\mathrm{N}_L \tilde{u}_\alpha + k_\alpha \mathrm{N}_{||} A_{||}(k). \tag{12.88}$$

The form factor $A_{||}(k)$ is associated with the gauge degrees of freedom and would be set to zero. It is easy to see that, this construction satisfies the Lorentz Gauge condition $k^\mu A_\mu = 0$. The equations of motion for the axions and photon form factors are given by,

$$\begin{aligned} \left(k^2 - \Pi_T(k)\right) A_2(k) - i\Pi_p N_1 N_2 \left[\epsilon_{\mu_\perp \nu_\perp 30} b^{(1)\nu} I^\mu\right] \mathrm{N}_1 A_1(k) &= -\frac{\left(iN_2 b^{(2)}_\mu I^\mu\right) a}{M_a}, \\ (k^2 - \Pi_T(k))A_1(k) + i\Pi_p N_1 N_2 \left[\epsilon_{\mu_\perp \nu_\perp 30} b^{(1)\mu} I^\nu\right] A_2(k) &= 0, \\ \left(k^2 - \Pi_L\right) A_L(k) &= \frac{iN_L \left(b^{(2)}_\mu \tilde{u}^\mu\right) a}{M_a}, \\ \left[\frac{\left(ib^{(2)}_\mu I^\mu\right)}{M_a} \mathrm{N}_2 A_2(k) + \frac{\left(ib^{(2)}_\mu \tilde{u}^\mu\right)}{M_a} \mathrm{N}_L A_L(k)\right] &= \left(k^2 - m^2\right) a. \end{aligned} \tag{12.89}$$

As in the previous case, in this case too we would assume the wave propagation to be in the z direction. and a generic solution written as $\Phi_i(t,z)$ for all the dynamical degrees of freedom would be assumed to be of the form, $\Phi_i(t,z) = e^{-i\omega t}\Phi_i(0,z)$. As we had done before, now we may express Eqs. (12.89), in real space in the matrix form

$$\left[(\omega^2 + \partial_z^2)\mathbf{I} - M\right] \begin{pmatrix} A_1(k) \\ A_2(k) \\ A_L(k) \\ a(k) \end{pmatrix} = 0. \tag{12.90}$$

where **I** is a $4 \times 4$ identity matrix and the modified mixing matrix, because of magnetized medium, turns out to be,

$$M = \begin{pmatrix} \Pi_T & -iN_1N_2\Pi_p\epsilon_{\mu_\perp\nu_\perp 30}b^{(1)\mu}I^\nu & 0 & 0 \\ iN_1N_2\Pi_p\epsilon_{\mu_\perp\nu_\perp 30}b^{(1)\nu}I^\mu & +\Pi_T & 0 & -i\frac{N_2b_\mu^{(2)}I^\mu}{M_a} \\ 0 & 0 & \Pi_L & -i\frac{N_Lb_\mu^{(2)}\tilde{u}^\mu}{M_a} \\ 0 & i\frac{N_2b_\mu^{(2)}I^\mu}{M_a} & i\frac{N_Lb_\mu^{(2)}\tilde{u}^\mu}{M_a} & m_a^2 \end{pmatrix}. \quad (12.91)$$

Solving this problem exactly is a difficult task, however in the low density limit one can usually ignore the effect of longitudinal field and $\Pi_L$. Again if we assume the $\omega \gg \omega_p$, then we can simplify the faraday contribution further. Incorporating these effects, the mixing matrix in this case turns out to be a $3 \times 3$ matrix, given by:

$$M = \begin{pmatrix} \omega_p^2 & i\omega_B\omega_p^2\cos\theta'/\omega & 0 \\ -i\omega_B\omega_p^2\cos\theta'/\omega & \omega_p^2 & -ig\mathcal{B}\omega \\ 0 & ig\mathcal{B}\omega & m_a^2 \end{pmatrix}, \quad (12.92)$$

The angle $\theta'$ is the angle between the magnetic field and the photon momentum $\vec{k}$, The other symbols are the same as used in the previously. This matrix can be diagonalized and one can obtain the exact result. The method of exact diagonalization of this matrix is relegated to the appendix.

The matrix given by eqn. [12.92] has been diagonalized and its eigen values have been evaluated perturbatively [(A. K. Ganguly P.K. Jain and S. Mandal , 2009)], in the limit $g\mathcal{B}\omega \gg \frac{\omega_B\omega_p^2\cos\theta'}{\omega} \gg |m_a^2 - \omega_p^2|$. The construction of the density (or coherency ) matrix from there is a straight forward exercise as illustrated before. Therefore instead of repeating the same here we would provide the values of the stokes parameters, computed from various components of the density matrix [(2)]. In this analysis we assume plane polarized light, with the following initial conditions $a(0) = 0$ and $A_1(0) = A_2(0) = \frac{1}{\sqrt{2}}$. That is the initial angle the beam makes with the direction of $I^\mu$ is $\pi/4$. The resulting stoke parameters are,

$$Q = -\sin(\Delta z), \qquad I = 1,$$

$$V = \frac{(g\mathcal{B})^2\,\omega^3\sin\left(\frac{\Delta z}{2}\right)\cos\left(\frac{\Delta z}{2} - \frac{\pi}{4}\right)}{\sqrt{2}\omega_B\,\omega_p^2\cos\theta'\,(m_a^2 - \omega_p^2)}, \qquad U = cos(\Delta z), \quad (12.93)$$

where in eqn. [12.93], the parameter $\Delta$ is given by, $\Delta = -2\frac{\omega_B\omega_p^2\cos\theta'}{\omega^2}$. Since V is associated with circular/ elliptic polarization, we can see from eqn. [12.93] that, even if one starts with a plane polarized wave, to begin with, it can become circularly or elliptically polarized light because of axion photon interaction and faraday effect. The ellipticity of the propagating wave turns out to be,

$$\chi = \frac{1}{2}\tan^{-1}\left(\frac{(g\mathcal{B})^2\,\omega^3\sin\left(\frac{\Delta z}{2}\right)\cos\left(\frac{\Delta z}{2} - \frac{\pi}{4}\right)}{\sqrt{2}\omega_B\,\omega_p^2\cos\theta'\,(m_a^2 - \omega_p^2)}\right). \quad (12.94)$$

---

(2) See for instance equation. [5.14], in [(A. K. Ganguly P.K. Jain and S. Mandal , 2009)]

and the polarization angle,$\psi$ would be given by,:

$$\tan(\psi + \pi/2) = -\cot(\Delta z). \tag{12.95}$$

when z is the path length traversed by the beam, in the magnetized media. We would like to emphasize here that, even in the limit of weak external magnetic field, it may not be prudent to ignore the contribution of Faraday effect. If we define a new energy scale $\omega_S$, such that

$$\omega_S = \left| \frac{\omega_B \left(\omega_p^4 - \omega_p^2 m_a^2\right) M^2 \cos\theta}{(\mathcal{B}^E)^2 \sin^2\theta} \right|, \tag{12.96}$$

then for $\omega_S \gg \omega$, to estimate $\chi$, one should consider the Faraday effect simultaneously.

We conclude here by noting that in this write up, we have tried to provide a comprehensive study of axion photon mixing and the associated observables of a photon beam. We have employed the coherency matrix formulation for studying the polarization properties; Starting with tree level axion photon interaction Lagrangian, we have demonstrated explicitly, how to construct the Stokes parameters from there. From there we have shown how to calculate the ellipticity angle and polarization angle from the Stokes Parameters. The relevant findings or questions pertaining to the current or proposed experiments in this area involve inclusion of matter effects, consideration of very strong magnetic field, dynamics of very high energy photon in such a scenario. Except the last, we have discussed the issues relevant for the first two. We end here by hoping that this elementary write up would help those who would like to take up advanced level investigations in this direction.

## 13. Acknowledgment

Many of the ideas I have presented here, took its shape during my collaborations with Prof. P. K. Jain and Dr. Subhayan Mandal. I am acknowledge them here in this note. I also would like to thank my wife, Dr. Archana Puri for her patience and understanding.

## 14. Appendix: Constructing the orthogonal matrix for diagonalization

Here we out line diagonalization of a $3 \times 3$ matrix given by eqn. (12.92), i.e., a symmetric matrix of the following type,

$$\mathbf{X_3} = \begin{bmatrix} a & b & 0 \\ b & c & d \\ 0 & d & g \end{bmatrix}. \tag{14.97}$$

Generalizing it to a hermitian matrix of the kind we have is trivial, so we would concentrate on diagonalizing the type given by eqn. (14.97). As noted already, the Cayley-Hamilton characterictic equation for this matrix looks like, $|\mathbf{X_3} - \lambda_\mathbf{i}| = 0$. for the i'th eigen value. Or for that matter, for any of the three eigen values, one should have:

$$\begin{vmatrix} a-\lambda_i & b & 0 \\ b & c-\lambda_i & d \\ 0 & d & g-\lambda_i \end{vmatrix} = 0 \tag{14.98}$$

Which when written in algebraic form looks like,

$$\lambda^3 - \lambda^2 (a + c + g) + \lambda \left(gc + ga + ac - d^2 - b^2\right) + \left(ad^2 + gb^2 - gac\right) = 0 \qquad (14.99)$$

Recalling that, the three roots of eqn. (14.99) satisfies the following relations

$$\begin{aligned} \lambda_1 + \lambda_2 + \lambda_3 &= (a + c + g) & (14.100)\\ \lambda_1\lambda_2 + \lambda_2\lambda_3 + \lambda_3\lambda_1 &= \left(gc + ga + ac - d^2 - b^2\right) & (14.101)\\ \lambda_1\lambda_2\lambda_3 &= -\left(ad^2 + gb^2 - gac\right) & (14.102) \end{aligned}$$

We should have for any value of $i(1, 2 or 3)$,

$$\begin{bmatrix} a-\lambda_i & b & 0 \\ b & c-\lambda_i & d \\ 0 & d & g-\lambda_i \end{bmatrix} \begin{pmatrix} u_i \\ v_i \\ w_i \end{pmatrix} = 0, \qquad (14.103)$$

with corresponding eigen-vector

$$\mathbf{V_i} = \begin{pmatrix} u_i \\ v_i \\ w_i \end{pmatrix}, \qquad (14.104)$$

All that we need to prove is ,

$$\mathbf{V_i} \cdot \mathbf{V_j} = \delta_{ij}. \qquad (14.105)$$

when suitably normalized. Next, assuming the eigen vectors to be normalized, we would demonstrate the necessary identities they need to satisfy. The proof should follow by explicit use of the values of $\lambda_i$'s in (14.105) (which is laborious ) or by some other less laborius method. Here we explore the last option. We write down the generic eqns. satisfied by the components of the eigen vectors

$$\begin{aligned} (a-\lambda)u + bv &= 0 \\ bu + (c-\lambda)v + dw &= 0 \\ dv + (g-\lambda)w &= 0. \end{aligned} \qquad (14.106)$$

It's easy to find out the nontrivial solns of (14.106) (for any of the three eigenvalues) by inspection and they are:

$$\begin{aligned} u &= -b(g-\lambda) \\ v &= (a-\lambda)(g-\lambda) \\ w &= -d(a-\lambda). \end{aligned} \qquad (14.107)$$

All that is to be shown is $\mathbf{V}_1 \cdot \mathbf{V}_2 = 0$ and other similar relations. We would prove the previous relation, others can be done using similar method. To begin with note that,

$$\begin{aligned} \mathbf{V}_1 \cdot \mathbf{V}_2 = &\Big[b^2(g-\lambda_1)(g-\lambda_2) + d^2(a-\lambda_1)(a-\lambda_2) \\ &+ (g-\lambda_1)(g-\lambda_2)(a-\lambda_1) \times (a-\lambda_2)\Big] \end{aligned} \qquad (14.108)$$

which is trivial to check. Next we start from,

$$[(g-\lambda_1)(g-\lambda_2)] = g^2 - g(\lambda_1+\lambda_2) + \lambda_1\lambda_2. \tag{14.109}$$

Eqn. (14.109) is a function of $\lambda_1$ and $\lambda_2$, and we need to convert it to a function of a single variable $\lambda_3$. To do that we would make use of the following tricks,

$$\begin{aligned}\lambda_1+\lambda_2 &= [\lambda_1+\lambda_2+\lambda_3] - \lambda_3 \\ \lambda_1\lambda_2 &= [\lambda_1\lambda_2+\lambda_2\lambda_3+\lambda_3\lambda_1] - \lambda_3(\lambda_2+\lambda_1)\end{aligned} \tag{14.110}$$

Now one can use the relations (14.101, 14.102 and 14.102), to replace the expressions inside the square bracket in eqns. (14.110) to get a function of only $\lambda_3$. i.e.

$$\begin{aligned}\lambda_1+\lambda_2 &= a+c+g-\lambda_3 \\ \lambda_1\lambda_2 &= gc+ga+ac-d^2-b^2-\lambda_3(a+c+g-\lambda_3).\end{aligned} \tag{14.111}$$

As one uses eqns. (14.111) in eqn. (14.109) one arrives at,

$$g^2 - g(\lambda_1+\lambda_2) + (\lambda_1.\lambda_2) = (\lambda_3-a)(\lambda_3-c) - b^2 - d^2. \tag{14.112}$$

so

$$b^2(g-\lambda_1)(g-\lambda_2) = b^2[(\lambda_3-a)(\lambda_3-c) - b^2 - d^2]. \tag{14.113}$$

Similarly one can show that,

$$d^2(a-\lambda_1)(a-\lambda_2) = d^2[(\lambda_3-g)(\lambda_3-c) - b^2 - d^2]. \tag{14.114}$$

Finally as we substitute in eqn. (14.108), the results of eqns. (14.113) and (14.114), we get after some cancellations,

$$\mathbf{V_1}\cdot\mathbf{V_2} = (c-\lambda_3)\left[(a-\lambda_3)(g-\lambda_3)(c-\lambda_3) - b^2(g-\lambda_3) - d^2(a-\lambda_3)\right] = 0, \tag{14.115}$$

because the expression inside the square bracket of eqn. (14.115) after the first = sign, is zero, as can be seen by expanding the determinant, i.e., eqn. (14.98) after taking $\lambda_i$ to be $\lambda_3$. In a similar fashion it can be shown that,

$$\mathbf{V_1V_2} = \mathbf{V_2V_3} = \mathbf{V_3V_1} = 0. \tag{14.116}$$

## 15. Proof: V's actually diagonalize the mixing matrix

Lets start from:

$$\begin{bmatrix} u_1 & u_2 & u_3 \\ v_1 & v_2 & v_3 \\ w_1 & w_2 & w_3 \end{bmatrix}^T \begin{pmatrix} a & b & 0 \\ b & c & d \\ 0 & d & g \end{pmatrix} \begin{bmatrix} u_1 & u_2 & u_3 \\ v_1 & v_2 & v_3 \\ w_1 & w_2 & w_3 \end{bmatrix} = \tag{15.117}$$

$$\begin{bmatrix} u_1a+bv_1 & u_1b+v_1c+w_1d & v_1d+gw_1 \\ u_2a+bv_2 & u_2b+v_2c+w_2d & v_2d+gw_2 \\ u_3a+bv_3 & u_3b+v_3c+w_3d & v_3d+gw_3 \end{bmatrix} \begin{pmatrix} u_1 & u_2 & u_3 \\ v_1 & v_2 & v_3 \\ w_1 & w_2 & w_3 \end{pmatrix}$$

Now if we recall (14.106), we see that,

$$\begin{aligned} au_1 + bv_1 &= \lambda_1 u_1 \\ bu_1 + cv_1 + dw_1 &= \lambda_1 v_1 \\ dv_1 + gw_1 &= \lambda_1 w_1. \end{aligned} \tag{15.118}$$

Similarly,

$$\begin{aligned} au_2 + bv_2 &= \lambda_2 u_2 \\ bu_2 + cv_2 + dw_2 &= \lambda_2 v_2 \\ dv_2 + gw_2 &= \lambda_2 w_2. \end{aligned} \tag{15.119}$$

And

$$\begin{aligned} au_3 + bv_3 &= \lambda_3 u_3 \\ bu_3 + cv_3 + dw_3 &= \lambda_3 v_3 \\ dv_3 + gw_3 &= \lambda_3 w_3. \end{aligned} \tag{15.120}$$

So we can substitute eqns. (15.118) to (15.120) in eqns. (15.118), to get:

$$\begin{bmatrix} u_1 a + bv_1 & u_1 b + v_1 c + w_1 d & v_1 d + gw_1 \\ u_2 a + bv_2 & u_2 b + v_2 c + w_2 d & v_2 d + gw_2 \\ u_3 a + bv_3 & u_3 b + v_3 c + w_3 d & v_3 d + gw_3 \end{bmatrix} \begin{pmatrix} u_1 & u_2 & u_3 \\ v_1 & v_2 & v_3 \\ w_1 & w_2 & w_3 \end{pmatrix} = \begin{bmatrix} u_1\lambda_1 & v_1\lambda_1 & w_1\lambda_1 \\ u_2\lambda_2 & v_2\lambda_2 & w_2\lambda_2 \\ u_3\lambda_3 & v_3\lambda_3 & w_3\lambda_3 \end{bmatrix} \tag{15.121}$$

$$\times \begin{pmatrix} u_1 & u_2 & u_3 \\ v_1 & v_2 & v_3 \\ w_1 & w_2 & w_3 \end{pmatrix} = \begin{bmatrix} \lambda_1 & 0 & 0 \\ 0 & \lambda_2 & 0 \\ 0 & 0 & \lambda_3 \end{bmatrix}$$

So we have checked that, the transformation matrix, constructed from the orthogonal vectors, diagonalize the mixing matrix.

## 16. References

Weinberg S. (1975). The U(1) problem. *Physical Review D* 11, 3583. (1975)

Belavin A.A., Polyakov A.M., Shvarts, A. S.and Tyupkin, yu. S. (1975). Inatanton Solutions In Nonabelian Gauge Theories. *Physics Letters B* 59, 85, (1975)

't Hooft, G. (1976). Symmetry Breaking Through Bell-Jackiw anomalies *Physical Review Letters,* Vol.37, 8, (1976).

't Hooft, G. (1976). Computation of the quantum effects due to a four dimensional pseudoparticle. *Physical Review D* 14, 3432 (1978); (E) ibid. 18, 2199, (1978).

R.J. Crewther (1978), "Effects of topological charge in gauge theories. *Acta Phys. Austriaca Suppl.* 19, 47.

V. Baluni(1979) "CP violating effects in QCD", *Physical Review D* 19, 2227.

R. J. Crewther, P. Di. Vecchia, G. Veneziano, E. Witten (1980) ,"Chiral estimates of the electric dipole moment of the neutron in quantum chromodynamics ", *Physics Letters B,* 88, 123; (E) ibid. B91, 487 (1980).

R. D. Peccei and H. R. Quinn (1977) , "CP Conservation in the Presence of Instantons," *Physical Review Letters,* Vol. 38, 1440. R. D. Peccei and H. R. Quinn(1977) , "Constraints

Imposed by CP Conservation in the Presence of Instantons," *Physical Review D* 16, 1791.

S. Weinberg (1978) , "A New Light Boson?," emphPhysical Review Letters, Vol 40, 223.

F. Wilczek (1978), "Problem of Strong p and t Invariance in the Presence of Instantons," *Physical Review Letters,* Vol 40, 279.

R.D. Peccei (1996), (*QCD, Strong CP and Axions*). *Journal of Korean Physics Society,* 29, S199. [arXiv:hep-ph/9606475]. For a more extensive review of the strong *CP* problem, see in *CP Violation*, edited by C. Jarlskog (World Scientific, Singapore, 1989).

M. S. Turner (1990) , "Windows on the Axion," *Physics Reports,* Vol 197, 67.

G. G. Raffelt (1990), "Astrophysical methods to constrain axions and other novel particle *Physics Reports* Vol 198, 1. G.G. Raffelt, in *Proceedings of Beyond the Desert*, edited by H.V. Klapder-Kleingrothaus and H. Paes (Institute of Physics, Bristol, 1998), p. 808.

G.G. Raffelt (1998) , in *Proceedings of 1997 European School of High-Energy Physics,* edited by N. Ellis and M. Neubert (CERN, Geneva, 1998), p. 235. Report No. hep-ph/9712538.

G.G. Raffelt (1996), *Stars as Laboratories for Fundamental Physics* (University of Chicago Press, Chicago, 1996).

J. Preskill M. B. Wise and F. Wilczek (1983) *Physics Letters,*B 120, 127.

M. Dine, W. Fischler and M. Srednicki (1981) , "A Simple Solution to the Strong CP Problem with a Harmless Axion," *Physics Letters,* B104, 199.

J. E. Kim (1979), "Weak Interaction Singlet and Strong CP Invariance," *Physical Review Letters,* Vol 43, 103.

L. M. Krauss, J. Moody and F. Wilczek (1985), *Physical Review. Letters,* Vol.55, 1797.

S . Moriyama, M. Minowa, T. Namba, Y. Inoue, Y. Takasu and A. Yamamoto (1998), "Direct search for solar axions by using strong magnetic field and X-ray detectors," *Physics Letters B* 434, 147. [hep-ex/9805026].

K. Zioutas et al.(2005) , [CAST Collaboration] , First results from the CERN axion solar telescope. *Physical Review Letters,* Vol 94, 121301.

M. Ruderman (1991), in *Neutron Stars: Theory and Observation*, edited by J. Ventura and D. Pines (Kluwer Academic, Dordrecht, 1991); G.S. Bisnovatyi-Kogan and S.G. Moiseenko (1992), *Astron. Zh.* Vol. 69, 563 (1992) [*Sov. Astron.* Vol. 36, 285 (1992)];

R.C. Duncan and C. Thompson (1992), *Astrophysical Journal,* 392, L9; C. Thompson and R.C. Duncan (1995) , *Mon. Not. R. Astron. Soc.* Vol. 275, 255 ; M. Bocquet (1995) *et al., Astron. Astrophys.* 301, 757.

A. K, Ganguly (2006),"Axion Photon Mixing in Magnetized media", *Annals Of Physics (N. Y.),* Vol 321, 1457.

J. S. Schwinger (1951), "On gauge invariance and vacuum polarization," *Physical Review,* Vol. 82, 664.

G. Raffelt and D. Seckel (1988), "Bounds on Exotic Particle Interactions from SN 1987a," *Physical Review Letters,* Vol. 60, 1793.

D. V. Galtsov and N. S. Nikitina (1972), "Photoneutrino processes in a strong field," *Zh. Eksp. Teor. Fiz.* Vol. 62, 2008.

L. L. . DeRaad, K. A. Milton and N. D. Hari Dass (1976), "Photon Decay Into Neutrinos in a Strong Magnetic Field," *Physical Review D* 14, 3326.

A. N. Ioannisian and G. G. Raffelt (1997), "Cherenkov radiation by massless neutrinos in a magnetic field," *Physical Review D* 55, 7038. [arXiv:hep-ph/9612285].

C. Schubert (2000), "Vacuum polarization tensors in constant electromagnetic fields. Part 2," *Nuclear Physics B* Vol. 585, 429. [arXiv:hep-ph/0002276].

A. K. Ganguly, S. Konar and P. B. Pal (1999), *Physical Review D* 60 105014. [arXiv:hep-ph/9905206].

K. Bhattacharya and A. K. Ganguly (2003) , "The Axialvector vector amplitude and neutrino effective charge in a magnetized medium," *Physical Review D* 68, 053011. [arXiv:hep-ph/0308063].

See for instance (A. K. Ganguly et al. , 1999).

M. Giovannini (2005), "Magnetized birefringence and CMB polarization," *Physical Review D* 71, 021301. [arXiv:hep-ph/0410387].

R. Cameron *et al.* (1993), "Search for nearly massless, weakly coupled particles by optical *Physical Review D* 47, 3707.

K. Bhattacharya and A. K. Ganguly (2003), "The Axialvector vector amplitude and neutrino effective charge in a *Physical Review D* 68, 053011. [arXiv:hep-ph/0308063].

R . Cameron. et al., (1993). Search for nearly massless, weakly coupled particles by optical techniques. *Physical Review D*, 47, 3707.

Avijit K. Ganguly, Pankaj Jain and Subhayan Mandal (2009): Photon and Axion Oscillation in a magnetized medium: A general treatment, *Physical Review D* 79, 115014.

# 4

# The e-Science Paradigm for Particle Physics[1]

Kihyeon Cho
*Korea Institute of Science and Technology Information*
*Republic of Korea*

## 1. Introduction

Research in the 21st century is increasingly driven by the analysis of large amounts of data within the e-Science paradigm. e-Science is the data centric analysis of science experiments unifying experiment, theory, and computing. According to Simon C. Lin and Eric Yen (Lin & Yen, 2009), e-Science or data-intensive science unifies theory, experiment, and simulations using exploration tools that link a network of scientists with their datasets. Results are analyzed using a shared computing infrastructure.

In this chapter, we use the concept of e-Science to combine experiment, theory and computing in particle physics in order to achieve a more efficient research process. Particle physics applications are generally regarded as a driver for developing this global e-Science infrastructure.

According to Tony Hey at Microsoft (Hey, 2006), thousands of years ago science focused on experiments to describe natural phenomena. In the last few hundreds of years, science became more theoretical. In the last few decades, science has become more computational, focusing on simulations. Today, science can be described as more data-intensive in nature, requiring a combination of experiment, theory, and computing. Attempts have been made to realize this e-Science concept. One e-Science application is the Worldwide Large Hadron Collider Computing Grid (WLCG), which realizes Ian Foster's definition of a grid (Foster et al., 2001). The grid is the combination of computing resources from multiple administrative domains to reach a common goal (Cho & Kim, 2009). As the global e-Science infrastructure is rapidly established, we must take advantage of worldwide e-Science progress. High-energy physics has advanced the e-Science paradigm by successfully unifying experiments, theory, and computing (Cho et al., 2011).

We apply the e-Science concept to particle physics and show an example of this paradigm. As shown in Fig. 1, we construct a unified research model of experiment-theory-computing in order to probe the Standard Model and search for new physics.

This is not a simple collection of experiments, computing, and theory, but a fusion of research in order to achieve a more efficient research process. We apply this concept to the

---

[1] This chapter is based on the paper titled "Collider physics based on e-Science paradigm of experiment-computing-theory" by K. Cho et al. in Computer Physics Communication Vol. 182, pp. 1756-1759 (2011).

Collider Detector at Fermilab (CDF) experiment in the USA and the Belle/Belle II experiment at High Energy Accelerator Research Organization (KEK) in Japan.

For computing-experiment, we construct and use the components of the e-Science research environment, including data production, data processing, and data analysis using collaborative tools. We also develop new computational tools for future experiments. In high energy physics, the goal of e-Science is to perform and/or analyze high energy physics experiments anytime and anywhere. We apply this system to the Belle II experiment at KEK. For data processing, WLCG is one of the original new research infrastructures that show how an effective collaboration might be conducted between users and facilities (Cho, 2007). The Asia Pacific area should develop both an e-Science platform and best practices for collaboration in order to fill the gaps in e-Science development between other continents. The Academia Sinica Grid Centre (ASGC), as the coordinator of the Asia federation under Enabling Grid in e-Science (EGEE), has worked closely with partners for region specific applications in data processing. For data analysis using collaborative tools, community building should be the foundation for collaboration rather than just offering technology. The e-Science research environment provides a trusted way to allow people, resources, and knowledge to connect and participate via a virtual organization. More and more countries will deploy a grid system and take part in the e-Science research environment. According to Simon C. Lin (Simon & Yen, 2009), we are widening the uptake of e-Science through close collaboration regionally and internationally.

For experiment-theory, we develop a combination of phenomenology and data analysis. Experiments give results and tools for theories and theories give feedback to experiments. We apply this system to the CDF, D0, and Belle experiments in order to probe the standard model and search for new physics. For theory-computing, we study lattice gauge theory and use the supercomputer at the Korea Institute of Science and Technology Information (KISTI).

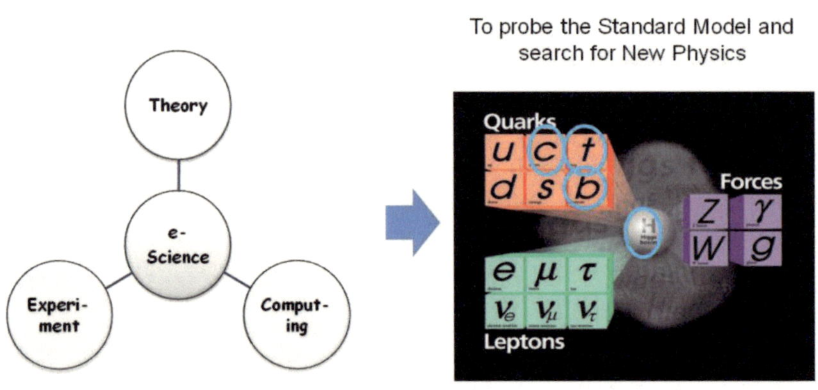


Fig. 1. The paradigm of e-Science in high energy physics, which is a fusion of experiment, computing, and theory research.

## 2. Main

We explain the results for computing-experiment, experiment-theory, and theory-computing for the analysis of particle physics. While many previous works have only used supercomputers, in our work computing results are combined with theory and experiment. We use a combination of supercomputers and an e-Science environment. The components of an e-Science environment are data production for remote shifts, data processing for grid farms, and data analysis using the Enabling Virtual Organization (EVO) collaborative tool.

### 2.1 For computing-experiment

#### 2.1.1 e-Science research environment

We define a computing-experiment tool as an e-Science research environment. In order to study particle physics, we can access the environment anytime and anywhere even if we are not on-site an accelerator laboratory. A virtual laboratory enables us to perform research as if we were on-site (Cho, 2008). We apply e-Science components to the CDF experiment.

**2.1.1.1 Data production**

The purpose of data production is to take both on-line shifts and off-line shifts anywhere. On-line shifts have been conducted through the use of a remote control room at KISTI and off-line shifts have been conducted via the sequential access through metadata (SAM) data handling (DH) system at KISTI. The remote control room is built to help non-US CDF members to fulfill their shift duties as a Consumer Operator (CO) part of the CDF data taking shift crew. The remote control room facilitates various monitoring applications that the CO has to monitor for a given eight hour shift. We have been operating the CDF remote control room at KISTI since July 22, 2008. A real Data Acquisition (DAQ) has been recorded at the remote control room at KISTI between August 1 and August 8, 2008. The CDF detector is an experimental apparatus for recording electrical events produced by the accelerator at an enormous rate. This apparatus is comprised of several components that perform different functions including a detector with millions of data channels transmitted to a corresponding number of electronic readout devices. The operation of an apparatus with this degree of complexity needs to be collaboratively controlled by researchers. In general, each shift crew takes an eight hour shift so that three shift crews will cover 24 hours. In the CDF experiment, the shift crew consists of three people with different missions. First, the Science Coordinator (SciCo) is responsible for the entire shift session and must have a lot of experience. The second person is the Ace shifter, who is an expert on the control of all detector components and electronic readout devices. The third person is the CO who has been trained in interpreting the meaning of the data being monitoring. UNIX processes intercept the on-line data transmitted from the front-end readout electronics and generate various plots that represent the quality of the data taken by the detector. These plots help the CO to determine whether or not the data collection is continuing as expected. Accordingly, the CO advises the Ace shifter to interrupt the detector operation in order to correct any problems.

Although the CO's monitoring task involves on-line data collection, this can be performed in a remote location due to its mostly monitoring-related nature. These remote control rooms are located at the Pisa University in Italy, the University of Tsukuba in Japan, and KISTI in Korea. In Korea, there are about 30 collaborators from six institutions, most of which have to fulfill CDF duties by taking detector operation shifts. All the plots that the consumers generate are accessible via web browsers where all the monitoring can be done. The CO has to not only monitor any plots generated by consumers but also must monitor the consumers themselves. However, the policy imposed by the Department of Energy (DOE) in the United States prohibits any remote researcher outside of Fermilab from executing any control-related UNIX command. Instead, control-related execution must be initiated by a person on-site. At the same time, all transmissions of control commands have to be encrypted using Kerberos. Thus, we can solve this problem by having an on-site crew send a graphic user interface (GUI) named "consumer controller" to the remote monitor via the Kerberized secure shell port. The CDF II experiment has been taking data from June 30, 2001 to September 30, 2011. Fig. 2 shows the CDF main operation center and remote control room at KISTI. As shown in Fig. 3, we have taken remote shifts (24 days per year on average) successfully.

Fig. 2. The CDF main operation center and remote control room at KISTI.

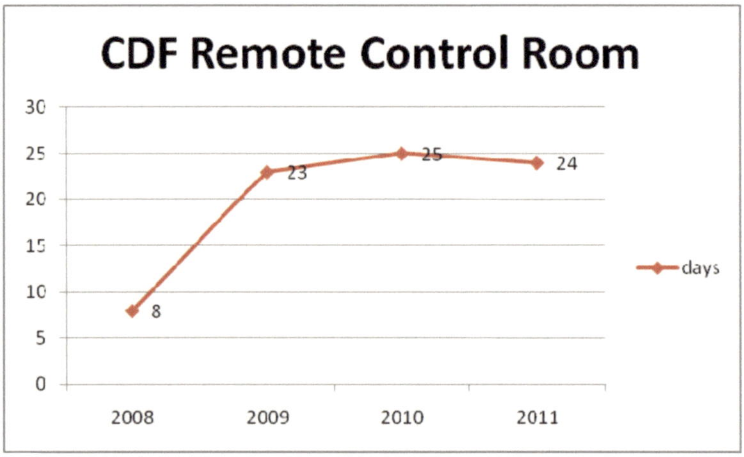


Fig. 3. The CDF remote control used at KISTI.

We perform another type of remote data handling shift at KISTI. Whereas the remote control room implements an on-line version of remote data handling, there is a second shift that implements an off-line version of remote data handling. This second type of shift is actually in the form of a SAM DH shift. This shift also occurs eight hours per day for seven days. These shifts do not need to cover the entire twenty four hours with three shifts per day since they are off-line. Furthermore, one can take the shift in the daytime of his or her time zone if participating in the shift schedule outside of the USA. The CDF SAM DH is called off-line since the data handled in this case includes data inbound to the tape from SAM stations in reconstruction farms and vice versa. The off-line data transfers in CDF are between SAM stations and mass storage system (MSS). In Fermilab, MSS consists of a Storage Resource Manager (SRM), dCache, and the Enstore system. The dCache software was the result of joint project between Fermilab in Batavia, USA and DESY (Deutches Elecktronen SYnchrotron laboratory) in Hamburg, Germany. dCache is a front-end for disk caching and provides end-users with the functionalities of reading cached files and writing files to and from Enstore indirectly via dCache. The Enstore system is a direct interface to files on tape for end-users. End-users can refer to SAM stations of CAF and farm machines. In the present context, the SAM stations in the CDF Analysis Farm (CAF) and farm clusters use an Application Programming Interface (API) provided by dCache to read files from and write files to the tapes via dCache and the Enstore systems. Thus, the mission of the CDF SAM shift includes monitoring the Enstore system, the dCache system, and SAM stations of the CDF analysis farm (CAF) and the CDF experiment farm.

#### 2.1.1.2 Data processing

Data processing is accomplished using a High-Energy Physics (HEP) data grid. The objective of the high-energy physics data grid is to construct a system to manage and process high-energy physics data and to support the high-energy physics community (Cho, 2007).

For data processing, Taiwan has the only WLCG Tier-1 center and Regional Operation Center in Asia since 2005. ASGC has also been serving as the Asia Pacific Regional Operational Center to maximize grid service availability and to facilitate extension of e-Science (Lin & Yen, 2009). In Japan, a Tier-2 computing center supporting the A Toroidal LHC Apparatus (ATLAS) experiment has been running at the University of Tokyo. There is another Tier-2 center at Hiroshima University for the A Large Ion Collider Experiment (ALICE) (Matsunaga, 2009). At KEK, collaborating institutes operate a grid site as members of the WLCG. These institutes try to use their grid resources for the Belle and Belle II experiments. The Belle II experiment, which will start in 2015, will use distributed computing resources.

We explain the history of data processing for the CDF experiment. The CDF is an experiment on the Tevatron, at Fermilab. The CDF group ran its Run II phase between 2001 and 2011. CDF computing needs include raw data reconstruction, data reduction, event simulation, and user analysis. Although very different in the amount of resources needed, they are all naturally parallel activities. The CDF computing model is based on the concept of a Central Analysis Farm. The increasing luminosity of the Tevatron collider has caused the computing requirement for data analysis and Monte Carlo production to grow larger than available dedicated CPU resources. In order to meet demand, CDF has examined the possibility of using shared computing resources. CDF is using several computing processing systems, such as CAF, Decentralized CDF Analysis Farm (DCAF), and grid systems. The

Korea group has built a DCAF for the first time. Finally, we have constructed a CDF grid farm at KISTI using an LCG farm.

In 2001, we have built a CAF, which is a cluster farm inside Fermilab in the United States. The CAF was developed as a portal. A set of daemons accept requests from the users via kerberized socket connections and a legacy protocol. Those requests are then converted into commands to the underlying batch system that does the real work. The CAF is a large farm of computers running Linux with access to the CDF data handling system and databases to allow the CDF collaborators to run batch analysis jobs. In order to submit jobs we use a CAF portal with two special features. First, we can submit jobs from anywhere. Second, job output can be sent directly to a desktop or stored on a CAF File Transfer Protocol (FTP) server for later retrieval (Jeung et al., 2009).

In 2003, we have built a DCAF, a cluster farm outside Fermilab. Therefore, CDF users around the world enabled to use it like CAF at Fermilab. A user could submit a job to the cluster either at Central Analysis Farm or at the DCAF. In order to run the remote data stored at Fermilab in USA, we used SAM. We used the same GUI used in Central Analysis Farm (Jeung et al., 2009).

In 2006, we have built CDF grid farms in North America, Europe, and Pacific Asia areas. The activity patterns at HEP required a change in the HEP computing model from clusters to a grid in order to meet required hardware resources. Dedicated Linux clusters on the Farm Batch System Next Generation (FBSNG) batch system were used when CAF launched in 2002. However, the CAF portal has gone from interfacing to a FBSNG-managed pool to Condor as a grid-based implementation since users do not need to learn new interfaces (Jeung et al., 2009).

We have now adapted and converted out a workflow to the grid. The goal of movement to a grid for the CDF experiment is a worldwide trend for HEP experiments. We must take advantage of global innovations and resources since CDF has a lot of data to be analyzed. The CAF portal may change the underlying batch system without changing the user interface. CDF used several batch systems. The North America CDF Analysis Farm and the Pacific CDF Analysis Farm is a Condor over Globus model, whereas the European CDF Analysis Farm is a LCG (Large Hadron Collider Computing Grid) Workload Management System (WMS) model. Table 1 summarizes the comparison of grid farms for CDF (Jeung et al., 2009). Fig. 4 shows the CDF grid farm scheme (Jeung et al., 2009). Users submit a job after they input the required information about the job into a kerberized client interface. The Condor over Globus model uses a virtual private Condor pool out of grid resources. A job containing Condor daemons is also known as a glide-in job. The advantage of this approach is that all grid infrastructures are hidden by the glide-ins. The LCG WMS model talks directly to the LCG WMS, also known as the Resource Broker. This model allows us to use grid sites where the Condor over Globus model would not work at all and is adequate for grid job needs. Since the Condor based grid farm is more flexible, we applied this method to the Pacific CDF Analysis Farm (Jeung et al., 2009).

The regional CDF Collaboration of Taiwan, Korea and Japanese groups have built the CDF Analysis Farm, which is based on grid farms. We called this federation of grid farms the Pacific CDF Analysis Farm.

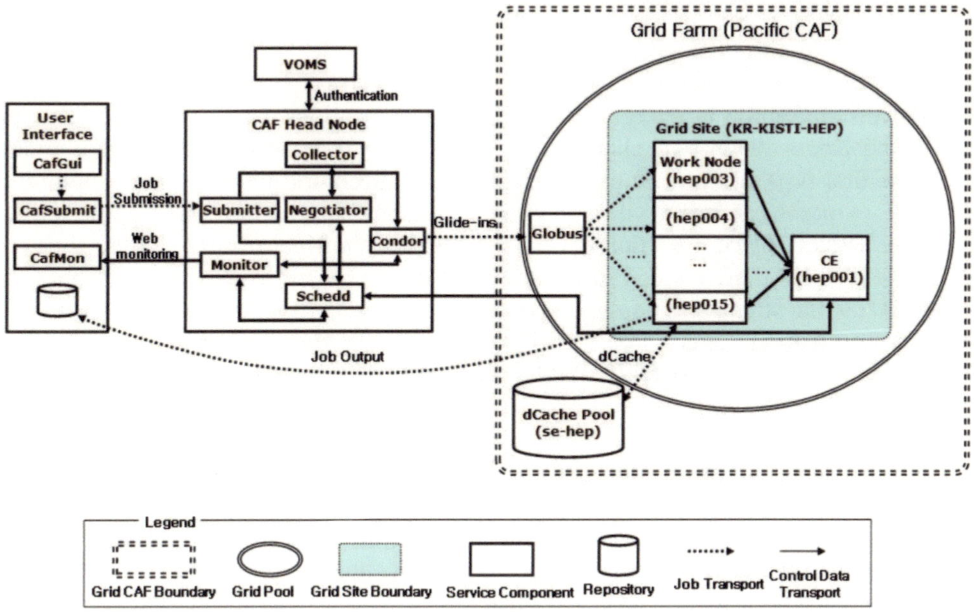


Fig. 4. The scheme of the Pacific CDF analysis farm.

| Grid CDF Analysis Farm | Head node | Work node | Grid middle-ware | Method | VO (Virtual Organization) |
|---|---|---|---|---|---|
| North America CDF Analysis Farm | Fermilab (USA) | USSD (USA) etc | OSG | Condor over Globus | CDF VO |
| European CDF Analysis Farm | CNAF (Italia) | IN2P3 (France) etc | LCG | WMS (Workload Management System | CDF VO |
| Pacific CDF Analysis Farm | AS (Taiwan) | KISTI (Korea) etc | LCG, OSG | Condor over Globus | CDF VO |

Table 1. Comparison of grid farms for CDF.

The Pacific CDF Analysis Farm is a distributed computing model on the grid. It is based on the Condor glide-in concept, where Condor daemons are submitted to the grid, effectively creating a virtual private batch pool. Thus, submitted jobs and results are integrated and are shared in grid sites. For work nodes, we use both LCG and Open Science Grid (OSG) farms. The head node of Pacific CDF Analysis Farm is located at the Academia Sinica in Taiwan. Now it has become a federation of one LCG farm at the KISTI in Korea, one LCG farm at the University of Tsukuba in Japan and one OSG and two LCG farms in Taiwan.

#### 2.1.1.3 Data analysis using collaborative tools

A data analysis using collaborative tools is for collaborations around the world to analyze and publish the results in collaborative environments. We installed an operator EVO server at KISTI. Using this environment, we study high energy physics for CDF and Belle experiments. EVO is the next version of its predecessor, Virtual Room Videoconferencing System (VRVS). The first release of EVO was announced in 2007. The EVO system is written in the Java programming language. The EVO system provides a client application named "Koala." The Koala plays two client roles in order to communicate with two types of servers. The first type is a central server located in Caltech and handles videoconferencing sessions. Participants can use a Koala to enter a session that another participant created or book a new session. Once a participant is in a session, the Koala starts to play the role of another type of client that now communicates with one of the networked servers that handle the flow of media streams. The second type of server comprising a network is called "Panda." When a Koala is connected to a specific Panda, the Koala initiates a video tool called "vievo" and an audio tool called "rat," both of which have their origins in the "MBone" project. EVO has improved upon VRVS with the following new features: support for Session Initiation Protocol (SIP), including ad-hoc or private meetings, encryption, private audio discussion inside a meeting, and whiteboard. In 2007, we constructed the EVO system at KISTI since the Korean HEP community is large enough to have its own EVO Panda servers. The configuration of two servers by the Caltech group enables the first Korean Panda servers to run. Fig. 5 shows communications between KISTI Panda servers and other Panda servers in the EVO network. Since its introduction in 2007, KISTI Panda servers have served many communities such as the Korean Belle community and the Korean CDF community.

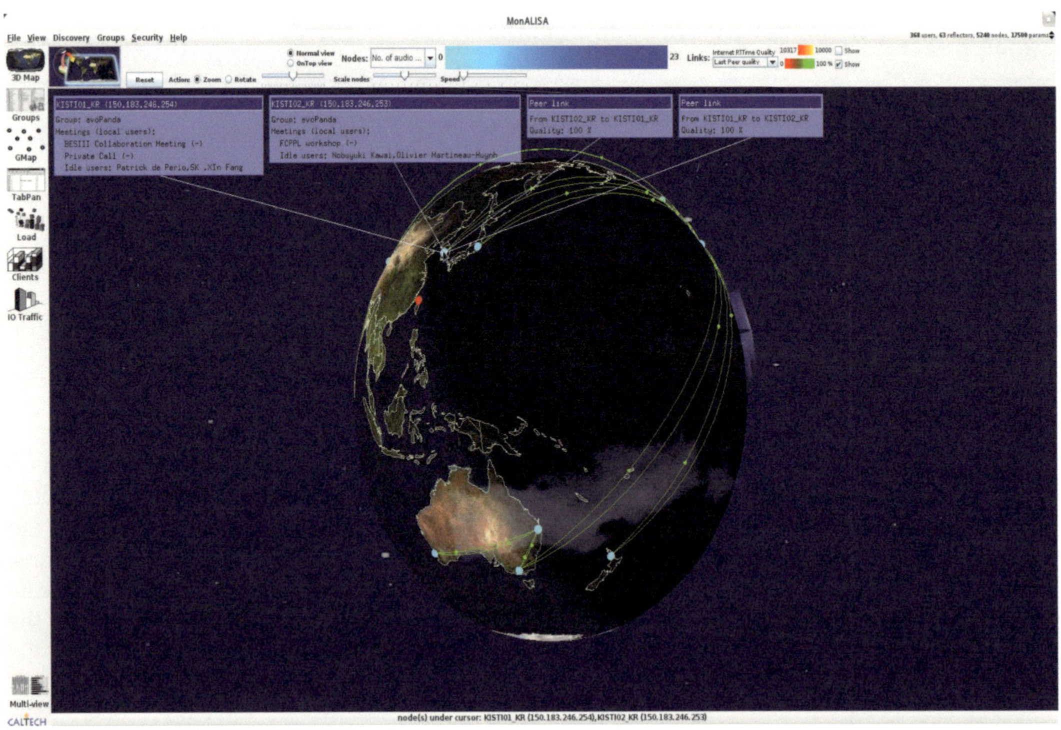


Fig. 5. Communications between KISTI "Panda" servers and other "Panda" servers in the EVO network.

### 2.1.2 New computing-experimental tools[2]

For new computing-experimental tools, we have worked on a Belle II data handling system. The Belle II experiment will begin at KEK in 2015. Belle II computing needs to include raw data reconstruction, data reduction, event simulation, and user analysis. The Belle II experiment will have a data sample about 50 times greater than that collected by the Belle experiment.

Therefore, we have very large disk space requirements and potentially unworkably long analysis times. Therefore, we suggested a meta-system at the event-level to meet both requirements. If we have good information at the meta-system level, we can reduce the CPU time required for analysis and save disk space.

The collider will cause the computing requirement for data analysis and Monte Carlo production to grow larger than available CPU resources. In order to meet these challenges, the Belle II experiment will use shared computing resources as the Large Hadron Collider (LHC) experiment has done. The Belle II experiment has adopted the distributed computing model with several computing processing systems such as grid farms (Kuhr, 2010).

In the Belle experiment (Abashian et al., 2002), we use a metadata scheme that employs a simple "index" file. This is a mechanism to locate events within a file based on predetermined analysis criteria. The index file is simply the location of interesting events within a larger data file. All these data files are stored on a large central server located at the KEK laboratory. However, for the Belle II experiment, this will not be sufficient as we will distribute the data to grid sites located around the world. Therefore, we need a new metadata service in order to construct the Belle II data handling system (Kim, et al. 2011; Ahn, et al., 2010).

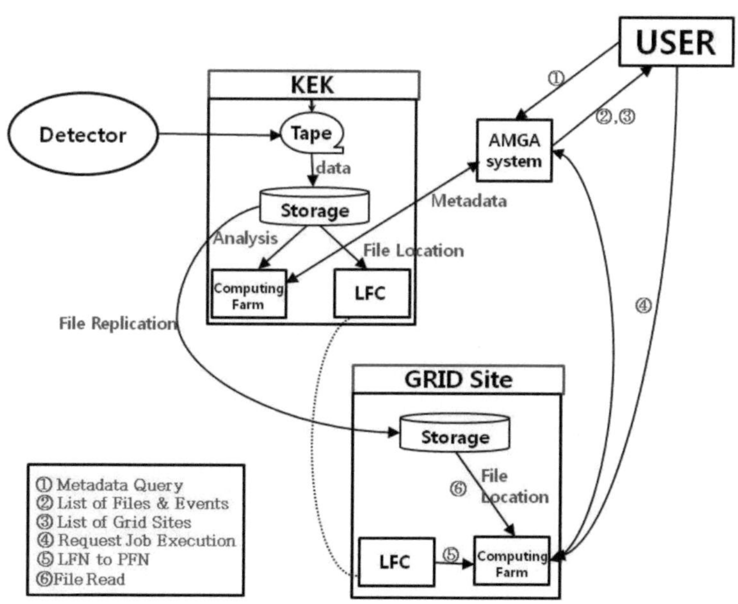


Fig. 6. Data handling scenario at the Belle II experiment.

[2] This section is based on the paper titled "The embedment of a metadata system at grid farms at the Belle II experiment" by S. Ahn et al. in Journal of the Korean Physical Society, Vol. 59, No. 4, pp. 2695-2701, (2011).

Fig. 6 shows the Belle II data handling system scheme. First, a user makes a metadata query to the server. Second, the server gives back a list of files and events. Third, the server may give a list of grid sites. Fourth, the user requests job execution at grid sites. Fifth, a logical file catalog (LFC) maps a logical file name (LFN) into a set of physical file names (PFN). Finally, the computing farms at the grid site read the requested physical file (Ahn, et al., 2011).

### 2.2 For experiment-theory

For experiment-theory, using the results of CDF and Belle experiments, we test phenomenological models of particle physics. Fig. 7 shows various physics topics for experiment-theory research, including Kaon Semi-leptonic form factor, rare B decay, mixing and *CP* (Charge Parity) violation on $Bs \rightarrow J/\psi\ \Phi$, forward-backward asymmetry of top quarks, and *CP* violating dimuon charge asymmetry due to B mixing. Models for these physics topics include lattice gauge theory using staggered fermion, Left-Right models, and model-independent analysis. In this section, we introduce the left-right model and the forward-backward asymmetry of top quarks

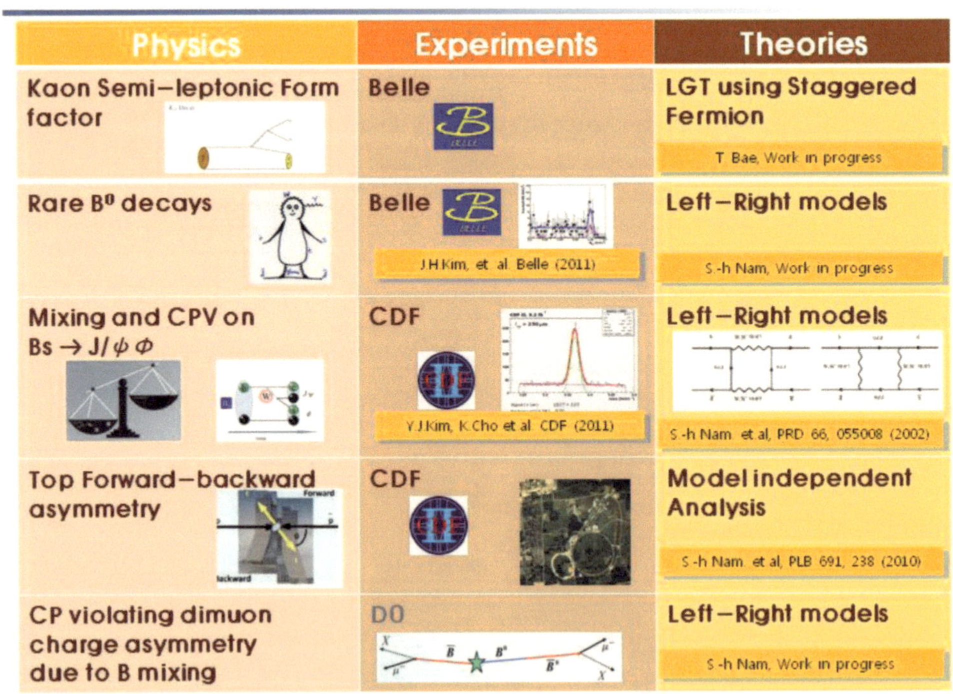


Fig. 7. Physics topics related to experiment and theory.

#### 2.2.1 Left- right models

In CDF experiments, we study mixing and *CP* violation on $Bs \rightarrow J/\psi\ \Phi$ decay channels. For this analysis, we apply Left-Right models and compare the results. We also apply to the same model to the *CP* violating dimuon charge asymmetry due to *B* mixing. Fig. 8 shows the

Feynman diagram of Left-Right models for the analysis of *CP* violating dimuon charge asymmetry due to *B* mixing.

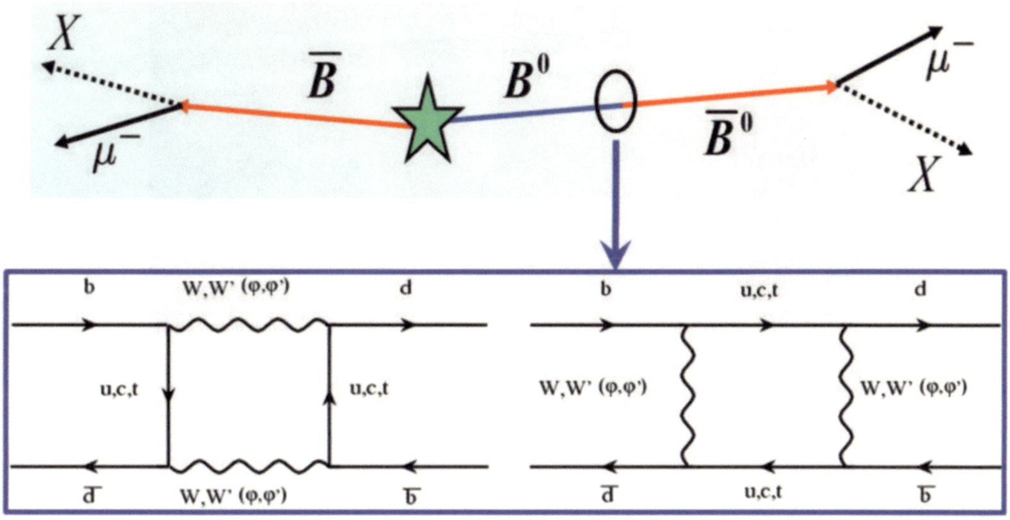


Fig. 8. The Feynman diagram of Left-Right models for the analysis of *CP* violating dimuon charge asymmetry due to B mixing.

### 2.2.2 The forward-backward asymmetry of top quark pairs

In 2008, CDF showed a possible anomaly in the forward-backward asymmetry of the top quark, where $A_{FB} = 0.19 \pm 0.07$(stat.) ± 0.02(syst.) (Aaltonen et al., 2008). We have performed model independent analysis. Considering the s-, t-, and u- channel exchanges of spin-0 and spin-1 particles whose color quantum number is a singlet, octet, triplet or sextet, we study the region consistent with the CDF data at a one sigma level. We show the necessary conditions for the underlying new physics in a compact and effective way when those new particles are too heavy to be produced at the Tevatron. However, the results still affect the forward-backward asymmetry of top quark.

### 2.3 For theory-computing

For theory-computing, we study flavor physics based on lattice gauge theory, which enables large-scale numerical simulations on a supercomputer. The theory of strong interactions in the Standard Model is Quantum Chromo Dynamics (QCD). In phenomena related to the Cabibbo-Kobayashi-Maskawa (CKM) matrix, the theoretical values of the interaction amplitudes also have factors that cannot be obtained in a perturbative way since the strong coupling constant becomes strong at a low energy scale as QCD, as a non-abelian gauge theory, predicts. The only way that one can calculate the non-perturbative quantities with a controlled error is the lattice method, in which we put strongly interacting particles, quarks and gluons, on a lattice and calculate quantities directly from first principles. Fig. 9 shows the baryon based on lattice QCD.

We use the staggered fermions, which are one of the more popular lattice fermion schemes for full QCD lattice simulations. The staggered fermion scheme has the advantage that its

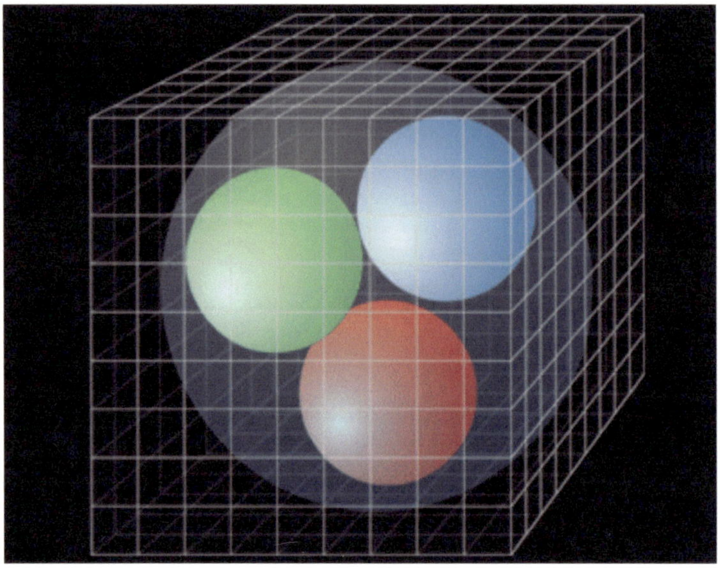

Fig. 9. Baryon based on lattice QCD.

computational cost is cheaper than other lattice fermion models while preserving remnant chiral symmetry. However, this scheme suffers from taste symmetry breaking in finite lattice spacing. Tastes are the remaining species that originate from the fermion doubling problem. Taste symmetry breaking complicates the analysis using lattice data. Thus, in order to reduce taste symmetry breaking effects, we use the HYP-smeared staggered fermions as valence quarks.

Lattice calculations cannot be done in the physical quark mass regime. In order to overcome this limitation, we calculate quantities with several non-physical quark masses and extrapolate the result to a physical regime. In this procedure, the staggered chiral perturbation theory guides the extrapolation.

This study can be extended to heavy flavor physics and other hadronic phenomena. In addition to physics research, we have developed new algorithms that enhance precision and utilize new hardware such as Graphic Processing Unit (GPU), which overcomes the limitation of CPU computing power.

### 2.3.1 Kaon semi-leptonic decay form factor

Fig. 10 shows the diagram for kaon semi-leptonic decay. The CKM matrix elements are quark mixing parameters, which can be determined by combining experimental weak decay widths of hadrons and their theoretical calculations. A traditional way to determine $V_{us}$ is connected with the kaon semi-leptonic decay channels, which include $K^+ \rightarrow \pi^0\ l^+\ \nu_l\ (K^+_{l3})$ and $K^0 \rightarrow \pi^-\ l^+\ \nu_l\ (K^0_{l3})$. Using these types of decays, we use the conserved vector current operator and the scalar density operator.

The decay rate of $K_{l3}$ is written as the product of $|V_{us}|^2$ and $|f_+(0)|^2$. The vector form factor at zero momentum transfer, $f_+(0)$, is defined from the hadronic matrix element of the vector current between kaon and pion states. The matrix elements of the vector current can be extracted from the three-point correlation function whose interpolating operators are composed by the pseudo-scalar operator and the conserved vector current operator.

In this method, we have to generate quark propagators first. In order to create the desired meson states (kaon or pion) with non-zero spatial momenta, we use random U(1) sources with momentum phases. We also use the PxP operator insertion method (generally called sequential source) in order to create or annihilate the other meson state. Next, we contract these quark propagators properly and obtain three-point correlation function data.

From a Ward identity, we can convert the matrix elements of the vector current operator to those of the scalar density operator. This gives another method to calculate the form factor. The way to obtain correlation function data is similar to that found for the vector current method. Since the two methods are connected by a Ward identity, we can check if the data is consistent.

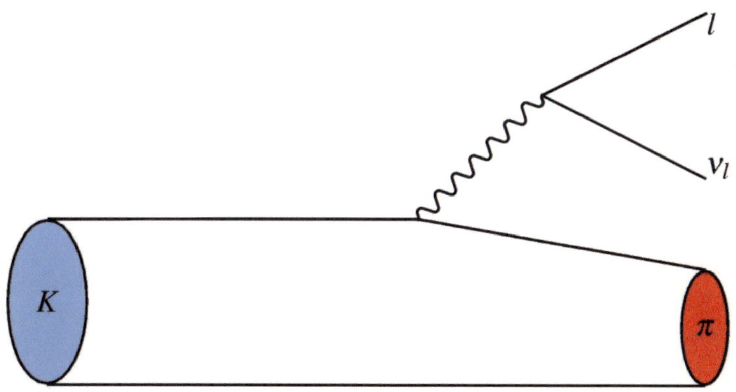


Fig. 10. Kaon semi-leptonic decay.

### 2.3.2 Kaon and pion decay constants

The kaon and pion decay constants can also be used to determine $V_{us}$. Since the ratio $f_K/f_\Pi$ is related to $V_{us}/V_{ub}$, we can obtain $V_{us}$ if $V_{ub}$ is precisely known. From these quantities, we calculate the two point function of axial vector current and pseudo-scalar operator in the same way as the form factor.

## 3. Conclusions

We have introduced the concept of an e-Science paradigm for experiment-computing-theory for particle physics. Computing-experiment collaborative research offers not only an e-Science research environment including data production, data processing and data analysis, but also a data handling system for the Belle II experiment. The e-Science research environment enables us to research particle physics anytime and anywhere in more efficient way. Experiment-theory collaborative research provides a way to study the standard model and new physics. Theory-Computing collaborative research enables lattice gauge theory tools using supercomputing at KISTI.

In conclusion, we presented a new realization of e-Science paradigm of experiment, theory and computing in particle physics. Applying this concept to particle physics, we can achieve more efficient results to test the standard model and search for new physics.

## 4. Acknowledgment

I would like to express my thanks to the members of high energy physics team (Junghyun Kim, Soo-hyeon Nam, Youngjin Kim and Taegil Bae) at KISTI for the work.

## 5. A glossary of acronyms

ALICE: A Large Ion Collider Experiment
API: Application Programming Interface
ASGC: Academia Sinica Grid Centre
ATLAS: A Toroidal LHC Apparatus
CAF: CDF Analysis Farm
CDF: Collider Detector at Fermilab
CKM: Cabibbo-Kobayashi-Maskawa
CO: Consumer Operator
CP: Charge-Parity
DAQ: Date Acquisition
DCAF: Decentralized CDF Analysis Farm
DESY: Deutches Elecktronen SYnchrotron laboratory
DH: Data Handling
DOE: Department of Energy
GUI: Graphic User Interface
EGEE: Enabling Grid in e-Science
EVO: Enabling Virtual Organization
FBSNG: Farm Batch System Next Generation
FTP: File Transfer Protocol
GPU: Graphic Processing Unit
HEP: High-Energy Physics
KEK: High Energy Accelerator Research Organization in Japan
KISTI: Korea Institute of Science and Technology Information
LCG: Large Hadron Collider Computing Grid
LFC: Logical File Catalog
LFN: Logical File Name
LHC: Large Hadron Collider
MSS: Mass Storage System
OSG: Open Science Grid
PFN: Physical File Name
QCD: Quantum Chromo Dynamics
SAM: Sequential Access through Metadata
SciCo: Science Coordinator
SIP: Session Initiation Protocol

SRM: Storage Resource Manager
VRVS: Virtual Room Videoconferencing System
WLCG: Worldwide Large Hadron Collider Computing Grid
WMS: Workload Management System

## 6. References

Aaltonen, T. et al. [CDF Collaboration] (2008). Forward-Backward Asymmetry in Top-Quark Production in ppbar Collisions at √s=1.96TeV, *Physical Review Letters*, Vol. 101, No. 20, pp. 202001, 0031-9007.

Abashian, A. *et al* [The Belle Collaboration] (2002). The Belle Detector, *Nuclear Instruments and Methods in Physics Research Section A: Accelerators, Spectrometers, Detectors and Associated Equipment*, Vol. 479, No. 1, pp. 117-232, 0168-9002.

Ahn. S., *et al.* (2010). Design of the Advanced Metadata Service System with AMGA for the Belle II Experiment, *Journal of Korean Physical Society*, Vol. 57, No. 4, pp. 715-724, 1976-8524.

Ahn, S., Kim, J. H., Huh, T., Hwang, S., Cho, K. *et al* (2011). The Embedment of a Metadata System at Grid Farms at the Belle II Experiment. *Journal of Korean Physical Society*, Vol. 59, No. 4, pp. 2695-2701, 1976-8524.

Cho, K. (2007). Cyberinfrastructure in Korea, *Computer Physics Communications*, Vol. 177, No. 1-2, pp 247-248, 0010-4655.

Cho K. (2008). e-Science for High Energy Physics in Korea, *Journal of Korean Physical Society*, Vol. 53, No. 92, pp.1187-1191, 1976-8524.

Cho, K. & Kim, H. W. (2009) Heavy Flavor Physics through e-Science, *Journal of Korean Physical Society*, Vol. 55, No. 52, pp. 2045-2050, 1976-8524.

Cho, K., Kim, H., & Jeung, M. (2010). Cyberinfrastructure for High Energy Physics in Korea, *Journal of Physics: Conference Series*, Vol. 219, No. 7, 072032, 1742-6596.

Cho, K., Kim, J. H., & Nam, S-H., (2011). Collider physics based on e-Science paradigm of experiment-computing-theory. *Computer Physics Communications*, Vol. 182, No. 9, pp. 1756-1759, 0010-4655.

Foster, I., Kesselman C., & Tuecke, S. (2001). The Anatomy of the Grid: Enabling Scalable Virtual Organizations, *International Journal of High-Performance Computing Applications*, Vol. 15, No. 3, pp. 200-222, 1094-3420.

Hey, T. (2006), e-Science and Cyberinfrastructure. Keynote lecture at the 20th International CODATA Conference, Beijing, China. 24 October 2006.

Jeung, M. *et al.* (2009). The Data Processing of e-Science for High Energy Physics, *Journal of Korean Physical Society*, Vol. 55, No. 52, pp 2067-2071, 1976-8524.

Kim, J. H. *et al.* (2011). The advanced data searching system with AMGA at the Belle II experiment, *Computer Physics Communications*, Vol. 182, No. 1, pp. 270-273, 0010-4655.

Kuhr, T. (2010). Computing at Belle II, *Proceedings of Computing in High Energy Physics 2010*, Taipei, October 2010.

Lin, S. C. & Yen, E. (2009). e-Science for High Energy Physics in Taiwan and Asia. *Journal of Korean Physical Society*, Vol. 55, No. 52, pp.2035-2039, 1976-8524.

Matsunaga, H. (2009). Grid Computing for High Energy Physics in Japan. *Journal of Korean Physical Society*, Vol. 55, No. 52, pp.2040-2044, 1976-8524.

# 5

# Muon Colliders and Neutrino Effective Doses

Joseph John Bevelacqua
*Bevelacqua Resources*
*USA*

## 1. Introduction

Lepton accelerators incorporate electron, muon, and tau beams. First generation lepton machines, electron accelerators, are basic research tools and their radiation characteristics are well established. A second generation muon machine presents additional research possibilities as well as new health physics challenges. Third generation tau accelerators are currently theoretical abstractions and little development has been forthcoming. Although this chapter focuses on muon colliders and their unique radiation characteristics, initial scoping calculatons for tau colliders are presented.

Neutrinos are electrically neutral particles, interact solely through the weak interaction, and have very small interaction cross sections (Particle Data Group 2010). They are present in the natural radiation environment due to cosmic rays, solar and terrestrial sources, and are produced during fission reactor and accelerator operations. From a health physics perspective these neutrino sources produce effective doses that are inconsequential. Although this will remain true for a number of years, planned muon accelerators or colliders will produce copious quantities of TeV energy neutrinos. In the TeV energy region, the health physics consequences of neutrinos can no longer be ignored. Upon operation of these accelerators, neutrino detection and the determination of neutrino effective doses will no longer be academic exercises, but will become practical health physics issues.

In a muon collider, neutrinos are produced when muons decay. The neutrino effective dose arises from neutrino interactions that produce showers or cascades of particles (e.g., neutrons, protons, pions, and muons). It is the particle showers that produce the dominant contribution to the neutrino effective dose (Bevelacqua, 2004).

Concerns for consequential neutrino effective doses have been previously postulated. Collar (1996) presented a hypothesis that the final stages of stellar collapse could produce neutrino effective doses that are sufficiently large to lead to the extinction of some species on earth. This concern has been challenged (Cossairt et al., 1997; Cossairt & Marshall, 1997), but the potential concern for large neutrino effective doses, on the order of hundreds of mSv/y or greater, remains, particularly for the planned muon colliders that will become operational in the next few decades of the 21$^{st}$ Century (Autin et al., 1999; Bevelacqua, 2004; Geer, 2010; King, 1999a; Kuno, 2009; and Zisman, 2011).

As background for muon colliders, an overview of the radiation environment at an electron accelerator is presented. This overview provides a foundation for a discussion of the characteristics of muon decays and the resultant neutrino effective doses. The characteristics of muon accelerators are addressed in this chapter and models for calculating the neutrino effective dose at a muon collider are provided. The radiological impacts of muon colliders and how basic dose reduction principles are affected by the underlying physics inherent in weak interaction processes are also discussed. Finally, a brief discussion of the neutrino effective doses anticipated at a third generation tau collider are provided.

## 2. Electron-positron colliders

Although this chapter addresses the neutrino effective dose from a muon collider, it is illustrative to provide a summary of the effects of other radiation types within a lepton collider (Bevelacqua, 2008, 2009, 2010a). These radiation fields are illustrated by considering an electron-positron collider. The radiation field within the muon collider facility are similar to those described in this section for electron-positron colliders.

An electron-positron collider accelerates electrons and positrons in circular rings before colliding the individual beams. There are a number of electron-positron colliders that have operated, are currently operating, or are being planned. These include the Large Electron Positron (LEP) Collider, and other machines summarized in the Review of Particle Properties (Particle Data Group, 2010). A new electron-positron machine, the International Linear Collider, is under design and is addressed from a health physics perspective in Bevelacqua (2008).

From an experimental physics perspective, electron-positron colliders have a number of advantages when compared to hadron colliders. First the collision results are less complex in terms of the particles produced, because electrons and positrons are fundamental particles without underlying structure or features. Hadrons are composed of quarks, but the electron and positrons have no such substructures. Therefore, the lepton's final state interactions are less complex than the structures that are produced from the interaction of the hadron's quarks. Particle interaction complexity is not the only advantage of electron-positron colliders.

The lepton colliders are also capable of achieving larger luminosities than hadron colliders. In addition, an order of magnitude less energy is required in electron-positron machines vice hadron colliders to achieve similar experimental results. For example, an electron-positron collider with a center-of-mass energy of 2 TeV is roughly equivalent to a 20 TeV center-of-mass energy hadron collider. In spite of these advantages, electron-positron collider health physics concerns exist (Bevelacqua; 2008, 2009, 2010a).

Electron-positron colliders produce more bremsstrahlung than hadron colliders. This bremsstrahlung production serves to limit the upper energies achieved by circular electron-positron colliders. In addition, electric power requirements rapidly increase with increasing energy unless beam power recovery mechanisms are developed and implemented.

The bremsstrahlung produced in a circular electron-positron collider is a fundamental concern that can only be decreased by increasing the circumference of the machine. The logical conclusion is to use an accelerator with an infinite radius (i.e., a linear collider). This is most easily achieved by replacing the dual beams in a circular collider with colliding beams from two linear colliders.

The electron and positron beams produce a variety of radiation types that are derived from the direct beam and its interactions. Secondary radiation is produced from bremsstrahlung when beam particles strike accelerator components and from synchrotron radiation when beam particles are defected by magnetic fields.

Bremsstrahlung has a number of health physics consequences. These health physics issues include (NCRP 144, 2003): (1) electromagnetic cascade radiation containing high-energy photons, electrons, and positrons, (2) high-energy radiation including neutrons, pions, muons, and other hadrons, (3) activation of accelerator structures and components, (4) activation of air, cooling water, and soil, and (5) ozone and oxides of nitrogen produced in the air. Synchrotron radiation also has health physics consequences including: (1) electromagnetic cascade radiation, (2) photons, (3) neutrons, (4) activation of accelerator structures and components, (5) activation of air, cooling water, and soil, and (6) ozone and oxides of nitrogen produced in the air. These secondary radiation categories and their health physics consequences are addressed in more detail in subsequent discussion and in Bevelacqua (2008, 2009, 2010a).

The primary electron (positron) beams are contained within beam tubes, and secondary radiation is produced when the primary particles exit the beam tube either by design or accident. When electrons (positrons) exit the beam tube they strike accelerator components such as the beam tube structure, vacuum components, collimators, or structural members. When this occurs, the beam particle decelerates and radiates photons through the process of bremsstrahlung. The high-energy, bremsstrahlung photons produce electron-positron pairs that lead to additional bremsstrahlung. This process repeats itself, and produces an electromagnetic shower or cascade that contains numerous particles and a spectrum of photons having energies up to the kinetic energy of the initial beam particles.

A second category of secondary radiation occurs when the beam particles traverse the accelerator's magnetic fields. The magnetic field produces a force that alters the particle's trajectory. It also changes the particle's velocity and leads to the emission of photon radiation. This process is known as synchrotron radiation. Synchrotron radiation is related to bremsstrahlung because a change in velocity or acceleration is involved in both processes. However, the synchrotron radiation differs from the bremsstrahlung spectrum.

With bremsstrahlung, the photon energy extends from zero up to the energy of the beam particle. However, synchrotron radiation is governed by the configuration and strength of the magnetic field. Therefore, the synchrotron spectrum is machine specific. For example, CERN's decommissioned Large Electron-Positron collider had a synchrotron spectrum that extended from the range of visible light to a maximum intensity that occurred in the range of a few hundred keV (Bevelacqua, 2008). The synchrotron radiation intensity rapidly decreases from its peak value as the photon energy increases above a few MeV. Both bremsstrahlung and synchrotron radiation induce an electromagnetic cascade.

The net result of the electromagnetic cascade is the deposition of energy in materials that are penetrated. This energy includes both particles stopped in the material and photon absorption. The photons produce additional secondary radiation and particles (e.g., photoneutrons) that activate accelerator materials. These same mechanisms lead to effective doses when personnel are in the presence of this radiation. These secondary radiation types are usually attenuated to insignificant levels by the concrete and earth shielding outside the accelerator tunnels containing the beam tubes.

From a health physics perspective, the energy loss of the circulating, accelerating electrons and positrons produces synchrotron radiation (photons). Given the mass of the electrons and positrons, their trajectories are easily altered. Therefore, synchrotron radiation is expected to be a large fraction of the available beam power. The synchrotron radiation requires shielding, and the extent of the shielding depends on the specific location within the accelerator facility.

The amount of synchrotron radiation depends on the specific design characteristics of the electron-positron collider. Dominant factors governing the production of synchrotron radiation are the beam power and radius of curvature of the accelerator ring. From a practical standpoint, radiation generated from the circulating electron and positron beams occurs within the unoccupied shielded ring and is not normally a health physics issue.

The dominant contributors to the radiation environment at an electron-positron facility include electromagnetic cascade showers, external bremsstrahlung, photoneutrons, muons, and synchrotron radiation. Muon pair production in the Coulomb field of a nucleus is possible above a photon energy of about 211 MeV. This process is analogous to electron-positron pair production, but the muon pair production cross-sections are smaller by a factor of about 40,000 due to the differences in electron (0.511 MeV) and muon (105.7 MeV) masses (Bevelacqua, 2008).

The dominant muon pair production process is coherent muon production. In coherent production, the target nucleus remains intact as it recoils from the photon interaction. In a few percent of the time, the nucleus breaks-up with the resultant emission of muons. Muons also result from the decay of photopions and photokaons. However, the number of muon decays in a conventional electron-positron collider is not sufficient to produce a neutrino effective dose concern. To understand the neutrino effective dose from a muon collider, it is necessary to understand neutrino physics and neutrino interactions.

## 3. Basic neutrino physics

The current view of elementary particle physics is embodied in the Standard Model of Particle Physics (Cottingham & Greenwood, 2007; and Griffiths, 2008) that assumes all matter is composed of three types of fundamental or elementary particles: leptons, quarks, and mediators of the fundamental interactions. Bevelacqua (2010b) provides a description of the Standard Model from a health physics perspective.

Leptons interact primarily through the weak interaction and electrically charged leptons also experience the effects of the electromagnetic force. They are not affected by the strong interaction. The leptons may be naturally grouped into three families or generations as $(e^-, \nu_e)$, $(\mu^-, \nu_\mu)$, and $(\tau^-, \nu_\tau)$.

Neutrinos are neutral leptons, once believed to be massless, but now evidence suggests they have a non-zero mass (Particle Data Group, 2010). The electron and muon neutrinos are well studied, but less is known about tau neutrinos

To allow for massive neutrinos, the Standard Model must be modified and its assumptions altered. However, current experimental knowledge of neutrino properties does not permit the selection of a specific modification to the model. For example, it is not known if neutrino masses are to be interpreted as evidence of new, light, fermionic degrees of freedom (e.g., Dirac neutrinos), new, heavy, degrees of freedom (e.g., Majorana neutrinos), or whether a more complicated electroweak-symmetry-breaking interaction is present. However, the Standard Model is sufficient for the purposes of this chapter.

Within the Standard Model, neutrino effective doses are determined from the muon decay processes:

$$\mu^- \rightarrow e^- + \nu_\mu + \bar{\nu}_e \tag{1}$$

$$\mu^+ \rightarrow e^+ + \bar{\nu}_\mu + \nu_e \tag{2}$$

The neutrino effective doses depend on the number of muon decays, and the subsequent production of neutrinos. Specific effective dose relationships are provided in subsequent discussion.

## 4. Neutrino interactions related to effective dose

In a muon collider, muon decays arise principally from Eqs. 1 and 2 that produce neutrinos and antineutrinos. The neutrinos interact through a variety of complex processes. A neutrino interaction discussion is simplified by following the methodology of Cossairt et al. (1997) and defining four processes (A, B, C, and D) to describe neutrino interactions with matter. The deposition of energy into tissue defines the effective dose (Bevelacqua, 2009, 2010a).

Process A involves neutrino scattering from atomic electrons. Electrons that recoil from elastic neutrino scattering deposit their energy in tissue and produce a neutrino effective dose. Process A occurs over a wide range of energy and the electron tissue interaction may involve multiple scattering of electrons.

In Process B, neutrinos interact coherently with nuclei. This process is only effective for low neutrino energies where the neutrino wavelength is too long to resolve the individual nucleons within the nucleus. At higher energies, Processes C and D become more important. Process B leads to low-energy ions having large linear energy transfer values. These ions deposit their energy into tissue according to their ranges, which are typically << 1 cm. Although Process B is independent of the neutrino generation, the cross section for neutrinos is about twice the antineutrino cross section (King 1999a).

Process C involves neutrino scattering from nucleons without shielding between the neutrinos and tissue. At energies below about 500 MeV, tissue dose is due to recoil nucleons. As the neutrino energy increases above about 0.5 GeV, secondary particle production increases. Eventually, these secondary particles produce particle showers or cascades in tissue. Process C is independent of the neutrino generation, affecting all three generations in the same manner.

Process D is similar to Process C with the exception that the neutrinos are shielded before striking tissue. Neutrinos with energy greater than about 0.5 GeV, emerging from a layer of material (e.g., earth shielding), result in a larger effective dose than unshielded neutrinos. The increase in effective dose arises from the fact that the tissue is exposed to the secondary particles produced by neutrino interactions in the shielding material as well as the neutrino beam. Process D is also independent of the neutrino generation.

A process that involves an increase in effective dose with added shielding is unique. One of the basic tenants for reducing effective dose for most radiation types (e.g., alpha and beta particles, heavy ions, muons, neutrons, photons, pions, and protons) is shielding the radiation source (Bevelacqua, 2009 and 2010a). The unique nature of Process D has a significant impact on the evaluation and control of neutrino effective dose.

## 5. Neutrino beam characteristics at a muon collider

Neutrinos are produced when the muon beam particles decay (See Eqs. 1 and 2). Weak interactions of muon neutrinos can be described in terms of two broad categories: charged current and weak current interactions. Charged current interactions involve the exchange of W-bosons to form secondary muons. Neutral current interactions produce uncharged particles through the exchange of Z-bosons. Both types of interactions produce hadron particle showers. Therefore, the neutrino induced radiation hazard will include secondary muons and hadronic showers. The hadronic showers have a much shorter range than the muons, but the number of particles in a hadronic shower can be quite large. The neutrino radiation hazard arises from these penetrating charged particle showers (Bevelacqua, 2008).

For TeV energy neutrinos, direct neutrino interactions in man account for less than 1% of the total effective dose because the primary hadrons from the neutrino interactions will typically exit the person before producing a charged particle shower (King, 1999b; Cossairt et al., 1996, 1997). Most of the neutrino effective dose is derived from particle showers produced in the shielding material.

The muon beam and subsequent neutrino beam are assumed to be well-collimated and to have a minimum divergence angle. For practical situations, the muons in the accelerator beam will have a small divergence angle and will be periodically focused using electromagnetic fields to ensure their collimation. No beam divergence is assumed in the subsequent calculations. Therefore, the actual beam will be somewhat more diffuse than assumed in the neutrino effective dose calculations. The neutrino beam will still produce particle showers, but they will be somewhat broader and less intense than the assumed well-collimated result. The beam divergence is analogous to the divergence of a laser beam as it exits an aperture (Bevelacqua, 2009, 2020).

The magnitude of the effective dose from a particle shower is dependent on the material in the interaction region lying directly upstream of the individual being irradiated. Calculation of the neutrino effective dose considers the configuration where a person is (1) completely bathed in the neutrino beam, and (2) is surrounded by material that will produce particle showers from neutrino interactions. These requirements lead to a bounding set of effective dose predictions.

These assumptions are too conservative for the TeV energies that will be encountered in mature muon colliders, but they provide a bounding neutrino effective dose result given the current level of design. Basic physics principles suggest that the neutrino interactions will be more peaked in the beam direction as the muon energies increase. In addition, the neutrino beam radius (r) will be relatively small and is given by (King, 1999b):

$$r = \theta L \tag{3}$$

where θ is called the characteristic angle, opening half-angle, or half-divergence angle of the muon decay cone

$$\theta = \frac{m\,c^2}{E} \tag{4}$$

In Eqs. 3 and 4, L is the distance to the point of interest such as the distance from the muon decay location to the earth's surface, θ is given in radians, E is the muon beam energy, and $mc^2$ is the rest mass of the muon (105.7 MeV). As the muon energy increases, the neutrino beam radius and size of the resultant hadronic showers tend to be smaller than the size of a person.

The characteristic angle varies inversely with energy. If E is expressed in TeV:

$$\theta \approx \frac{10^{-4}}{E[TeV]} \tag{5}$$

Therefore, the emergent neutrino beam will consist of a narrow diverging beam that is conical in shape.

Table 1 summarizes straw-man muon collider parameters (King, 1999b). It should be noted that the straw-man muon colliders are constructed below the earth's surface to provide muon shielding. However, the neutrino attenuation length is too long for the beam to be appreciably attenuated by any practical amount of shielding, including the expanse of ground between the collider and its exit from the surface of the earth. Therefore, the effective dose reduction principle as applied to neutrinos will no longer include shielding as an element. In fact, shielding the neutrino beam will produce hadronic showers and increase the effective dose. This peculiar behavior has its basis in the nature of the weak interaction, the uncharged nature of the neutrino, and the TeV energies that will be encountered in proposed muon colliders.

| E (TeV) | 2 | 5 | 50 |
|---|---|---|---|
| L (km) | 62 | 36 | 36 |
| r (m) | 3.3 | 0.8 | 0.08 |
| Collider depth (m) | 300 | 100 | 100 |

Table 1. Straw-Man Muon Collider Parameters.

The neutrinos exiting a muon collider will not only have a narrow conical shape, but will also have an extent that is quite long. The long, narrow plume of neutrinos will produce secondary muons and hadronic showers at a significant distance from the muon collider. This distance will be greater than tens of kilometers for TeV muon energies.

## 6. Neutrino interaction model

Neutrinos can interact directly with tissue or with intervening matter to produce charged particles that result in a biological detriment. The radiation environment is complex and simulations (e.g., Monte Carlo methods) can be used to model the dynamics of the neutrino interaction including the energy and angular dependence of each particle (e.g., $\nu_e$, $\bar{\nu}_e$, $\nu_\mu$, $\bar{\nu}_\mu$, $\nu_\tau$, $\bar{\nu}_\tau$, e, μ, τ, and hadrons) involved in the interaction. Performing a neutrino simulation is too dependent on specific accelerator characteristics and will not add to the health physics presentation. Rather than performing a Monte Carlo simulation, we follow the analytical approach of Cossairt et al. (1997) and King (1999b) to quantify the neutrino effective dose. This approach is acceptable in view of the current uncertainties in muon collider technology and the nature of the neutrino interaction for both charged current (CC) and neutral current (NC) weak processes (King, 1999c).

Following King (1999c), the dominant interaction of TeV-scale neutrinos is deep inelastic scattering with nucleons that include CC and NC components. In the NC process, the neutrino is scattered by a nucleon (N) and loses energy with the production of hadrons (X) through a $\nu+N\rightarrow\nu+X$ reaction. This NC reaction contributes about 25 percent of the total cross section. This NC process can be interpreted as elastic scattering off one of the quarks (q) inside the nucleon through the exchange of a virtual $Z^0$ boson ($\nu+q\rightarrow\nu+q$).

CC scattering is similar to NC scattering except that the neutrino is converted into its corresponding charged lepton (l). This includes reactions such as $\nu+N\rightarrow l^-+X$ and $\bar{\nu}+N\rightarrow l^++X$ where l is an electron/muon for electron/muon neutrinos. At the quark level, a charged W boson is exchanged with a quark to produce another quark (q') whose charge differs by one unit through processes such as $\nu+q\rightarrow l^-+q'$ and $\bar{\nu}+q'\rightarrow l^++q$.

The final state quarks produce hadrons on a nuclear distance scale that contribute to the effective dose. The CC and NC scattering processes are included in the Process A -D descriptions noted in previous discussion.

## 7. Neutrino effective dose

A muon collider provides a platform for colliding beams of muons ($\mu^-$) and antimuons ($\mu^+$) (Geer, 2010). The collider may involve a pair of linear accelerators with intersecting beams or a storage ring that circulates the muons and antimuons in opposite directions prior to colliding the two beams. The accelerator facility energy is usually expressed as the sum of the muon and antimuon energies. For example, a 100 TeV accelerator consists of a 50 TeV muon beam and a 50 TeV antimuon beam. Since muon colliders produce large muon currents, neutrinos will be copiously produced from the decay of both muons and antimuons (See Eqs. 1 and 2).

Neutrino effective dose calculations are performed for two potential muon collider configurations. The first configuration utilizes the intersection of the beams of two muon linear colliders. The linear collider effective dose model incorporates an explicit representation of the neutrino cross section and evaluates the effective dose assuming specific values for the muon energy, number of muon decays per year, and accelerator operational characteristics (e.g., accelerator gradient or the increase in muon energy per unit accelerator length). The operational parameter approach is more familiar to high-energy physicists, but it serves to illustrate the sensitivity of the neutrino effective dose to the key muon collider's operating parameters.

The second configuration is a circular muon collider. The neutrino effective dose for the circular muon collider involves an integral over energy of the differential fluence and fluence to dose conversion factor. This approach is more familiar to health physicists, but much of the muon collider's operating parameters are absorbed into other parameters and are not explicitly apparent. Using both approaches yields not only the desired neutrino effective dose, but also illustrates the sensitivity of the effective dose to a number of accelerator parameters and operational assumptions.

### 7.1 Bounding neutrino effective dose – linear muon collider

The bounding neutrino effective dose from a linear muon collider is derived following King (1999b) and is based on the effective dose from a straight section (ss) of a circular muon collider. This derivation incorporates a limiting condition from a circular accelerator with a number of straight sections as part of the facility. Parameters unique to the circular collider such as the ring circumference and straight section length appear in intermediate equations, but cancel in the final effective dose result. In the linear muon collider, the muon beam is assumed to be well-collimated.

In a linear muon collider, the total neutrino effective dose (H) is defined in terms of an effective dose contribution δH(E) received in each energy interval E to E + dE as the muons accelerate to the beam energy $E_o$:

$$H = \int_0^{E_o} dE\,\delta H(E) \tag{6}$$

The effective dose contribution δH(E) is written as (King, 1999b):

$$\delta H(E) = H' \frac{1}{f_{ss}} \frac{df(E)}{dE} \tag{7}$$

where $\frac{df(E)}{dE} dE$ is the fraction of muons that decay via Eqs. 1 and 2 in the energy interval E to E + dE, which may be written as:

$$\frac{df(E)}{dE} = \frac{1}{\gamma \beta c \tau g} \tag{8}$$

where

$$\gamma = \frac{E_o}{mc^2} \tag{9}$$

In Eq. 8, $\beta = v / c$, $\tau$ is the muon mean lifetime ($2.2 \times 10^{-6}$ s), and g is the accelerator gradient (dE/dl). The other parameters appearing in Eq. 7 include $f_{ss}$ (the ratio of the straight section length to the ring circumference) and H' (the effective dose that is applicable as the muon energy reaches the TeV energy range), where

$$f_{ss} = \frac{l_{ss}}{C} \tag{10}$$

In Eq. 10, C is the ring circumference:

$$C = \frac{2\pi E_o}{0.3\overline{B}} \tag{11}$$

In Eqs. 9 - 11, v is the muon velocity, $l_{ss}$ is the straight section length, $E_o$ is the muon energy, $\overline{B}$ is the ring's average magnetic induction, and N is the number of muon decays in a year.

In the narrow beam approximation, the effective dose is independent of distance (L) for L < 5 $E_o$ (King, 1999b) where L is expressed in km and $E_o$ in TeV. Using this approximation,

$$H' = K' \, N \, l_{ss} \, \overline{B} E X \tag{12}$$

where K' is a constant that depends on the units used to express the various quantities appearing in Eq. 12, and X = X(E ) is the cross section factor defined in subsequent discussion.

Combining these results leads to the annual neutrino effective dose (H) in mSv/y:

$$H = \frac{N K}{g} \int_{0}^{E_o} E X(E) dE \tag{13}$$

where K = $6.7 \times 10^{-21}$ mSv-GeV /m-TeV$^2$ if g is expressed in GeV/m, N is expressed in muon decays per year, E is the muon energy in TeV, and the cross section factor is dimensionless (Bevelacqua, 2004).

In deriving the linear muon collider effective dose relationship, a number of assumptions were made (Bevelacqua, 2004). These assumptions are explicitly listed to ensure the reader clearly understands the basis for Eq. 13. The relevant assumptions include applicability of the narrow beam approximation. The individual receiving the effective dose is assumed to be: (1) uniformly irradiated, (2) within the footprint of the neutrino beam, (3) within the footprint of the hadronic particle shower that results from the neutrino interactions, and (4) irradiated by only one of the linear muon accelerators whose energy is one-half the total linear muon collider energy. Given the TeV muon energies and the earth shielding present, charged particle equilibrium exists and Process D dominates the neutrino effective dose. In addition, the muon beam is well-collimated, the neutrino effective dose calculation assumes a 100% occupancy factor, and the neutrino effective dose is an annual average based on the number of muon decays in a year.

The cross section factor is a parameterization of the neutrino cross section (See Table 2) in terms of a logarithmic energy interpolation (Quigg, 1997). The numerical factors in the Table

2 expressions (1.453, 1.323, 1.029, 0.512, and 0.175) are the total summed neutrino-nucleon and antineutrino-nucleon cross sections divided by energy at neutrino energies of 0.1, 1, 10, 100, and 1000 TeV, respectively, given in units of $10^{-38}$ $cm^2/GeV$. As an approximation, the muon energies in Table 2 are set equal to the corresponding neutrino energies. Following Quigg (1997), the cross section factor is a dimensionless number and is normalized such that X(E = 0.1 TeV) = 1.0.

| Muon Energy Range (TeV) | X(E) |
|---|---|
| E < 1 | $(-1.453\ \alpha + 1.323\ (\alpha + 1))\ /\ 1.453$ |
| 1 < E < 10 | $(1.323\ (1-\alpha) + 1.029\ \alpha)\ /\ 1.453$ |
| 10 < E < 100 | $(1.029\ (2-\alpha) + 0.512\ (\alpha-1))\ /\ 1.453$ |
| 100 < E < 1,000 | $(0.512\ (3-\alpha) + 0.175\ (\alpha-2))\ /\ 1.453$ |
| E > 1,000 | $(0.175/1.453)\ 3^{3-\alpha}$ |
| $\alpha = \log_{10}(E)$ where E is the muon energy expressed in TeV. | |

Table 2. Cross Section Factor X(E) as a Function of Muon Energy.

Eq. 13 may be approximated by replacing the energy-weighted integral of X(E) by its value at $E = E_o/2$. This choice is acceptable given the energy dependence of the cross section and the associated uncertainties in the collider design parameters. With this selection, the annual neutrino effective dose (mSv/y) becomes:

$$H = \frac{KN}{2g} X(E_o/2)\, E_o^2 \tag{14}$$

As a practical example (Zimmerman, 1999), consider a 1,000 TeV muon linear accelerator assuming $E_o$ = 500 TeV (i.e., two, 500 TeV linear muon accelerators) and N = 6.4 x $10^{18}$ muon decays per year. Using these values in Eq. 14 with a g = 1 GeV/m value leads to an annual effective neutrino dose of 1.4 Sv/y, which is a significant value that cannot be ignored. Health physicists at a linear muon collider will need to contend with large neutrino effective doses within and outside the facility. Table 3 provides expected annual neutrino effective doses for a variety of accelerator energies using the same N and g values noted above and the narrow beam approximation.

| Accelerator Facility Energy (TeV) | Muon Beam Energy (TeV) | H (mSv/y) |
|---|---|---|
| 0.1 | 0.05 | $5.7 \times 10^{-5}$ |
| 1 | 0.5 | $5.2 \times 10^{-3}$ |
| 10 | 5 | 0.45 |
| 100 | 50 | 30 |
| 500 | 250 | 440 |
| 1,000 | 500 | $1.4 \times 10^{3}$ |
| 5,000 | 2,500 | $1.5 \times 10^{4}$ |
| 10,000 | 5,000 | $4.2 \times 10^{4}$ |
| 50,000 | 25,000 | $4.8 \times 10^{5}$ |

Table 3. Annual Neutrino Effective Doses for a Linear Muon Collider Using the Narrow Beam Approximation.

The values of Table 3 suggest that the annual effective dose limit for occupational exposures of 20 mSv/y and the annual effective dose limit to the public (1 mSv/y) can be exceeded by TeV energy muon accelerators (ICRP 103, 2007). The values in Table 3 also exceed the emergency effective dose limit of 250 mSv set for the Fukushima Daiichi accident that is based on ICRP 60 (1991).

A TeV - PeV scale muon collider will also challenge the acute lethal radiation dose ($LD_{50,30}$) of about 4 Gy (Bevelacqua 2010a). Although the feasibility of TeV - PeV scale machines remains to be determined, the significant radiation hazards associated with their operation merits careful attention to the effects of neutrino effective doses at offsite locations.

Selecting an accelerator location will be an issue for TeV energy muon linear colliders due to public radiation concerns arising from neutrino interactions. Given these radiation concerns, a muon collider location may be restricted to low population or geographically isolated areas to minimize the public neutrino effective dose.

### 7.2 Bounding neutrino effective dose – circular muon collider

The bounding neutrino effective dose for a circular muon collider could be obtained using the methodology of the previous section. However, a number of operational assumptions including the ring circumference and average magnetic induction would be required. Instead, we use an alternative approach to illustrate the various methods than can be utilized to determine the neutrino effective dose as a function of distance. To accomplish this, consider the energy distribution or differential fluence $dN_i(E_i)/dE_i$ where $N_i$ is the number of neutrinos of generation i per unit area, $E_i$ is the neutrino energy, and i = 1, 2, and 3 for the three neutrino generations. The neutrino effective dose H can be determined once the neutrino fluence to effective dose conversion factor $C(E_i)$ is known.

Cossairt et al. (1997) provide an approach for treating the neutrinos and their antiparticles in the first two generations. In view of the limited data, Cossairt et al. (1997) did not consider the generation 3 neutrinos, but these neutrinos become more important as the accelerator energy increases.

One of the initial goals of a muon accelerator will be the development of a pure muon neutrino beam to investigate the magnitude of the neutrino mass. Focusing on the muon neutrino is also warranted because Cossairt et al. (1997) provides a muon neutrino fluence to effective dose conversion factor. Following Cossairt et al. (1997) and Silari & Vincke (2002), we limit the subsequent discussion to muon neutrinos that result from muon decays (Eq. 1) in a circular muon collider and drop the subscript i:

$$H = \int_0^{E_o} \frac{dN(E)}{dE} C(E) dE \tag{15}$$

where $E_o$ is the energy of the primary muons before decay.

Silari & Vincke (2002) provide a differential fluence value in the laboratory system that is averaged over all neutrino production angles. They also assume the accelerator's shielding is thick enough to attenuate the primary muon beam, and that it is thicker than the range of all secondary radiation. Accordingly, the neutrino radiation is in equilibrium with its secondary radiation.

Using the equilibrium condition and averaging over all production angles, provides the following differential fluence relationship for the neutrino radiation from a circular muon collider (Silari & Vincke, 2002):

$$\frac{dN(E)}{dE}=\frac{2}{E_o}\left(1-\frac{E}{E_o}\right)\Phi \tag{16}$$

where N(E) is the number of neutrinos per unit area, E is the neutrino energy, $E_o$ is the energy of the primary muons before decay, and $\Phi$ is the integral neutrino fluence (total number of neutrinos per unit area) following the muon decays.

For secondary particle equilibrium, the fluence to effective dose conversion factor relationship of Cossairt et al. (1997) is used:

$$C(E)=KE^2 \tag{17}$$

Eq. 17 was derived for the neutrino energy range of 0.5 GeV to 10 TeV. In deriving the muon neutrino effective dose to fluence conversion factor of Eq. 17, Cossairt et al. (1997) did not consider the effects of the third lepton generation.

In Eq. 17, K = $10^{-15}$ µSv-cm$^2$/GeV$^2$. In view of the trend in the neutrino data (Particle Data Group, 2010; Quigg, 1997), Eq. 17 is used at energies beyond those considered by Cossairt et al. (1997). This is reasonable because increasing energy and increasing number of secondary shower particles (hadrons) is the main reason for the rising fluence to effective dose conversion factor with increasing neutrino energy for the equilibrium (shielded neutrino) case or process D described earlier. It is also reasonable because the neutrino attenuation length ($\lambda$) decreases with increasing energy of the primary neutrinos. Although TeV energy units are used in the final result, GeV units are used in the derivation of the neutrino effective dose to facilitate comparison with Silari & Vincke (2002) and Johnson et al. (1998). Prior to developing the neutrino effective dose relationship for a circular muon collider, the neutrino attenuation length is briefly examined.

The neutrino attenuation length is written in terms of the neutrino interaction cross section $\sigma_v$:

$$\lambda=\frac{A}{\rho N_A \sigma_v}=\frac{1}{N\sigma_v} \tag{18}$$

where A and $\rho$ are the atomic number and density of the shielding medium, $N_A$ is Avogadro's number, N is the number density of atoms of the shielding medium per unit volume, and $\sigma_v$ is on the order of $10^{-35}$ cm$^2$ (E / 1 TeV ) (Johnson et al. ,1998) where the neutrino energy is expressed in TeV.

These results permit the neutrino attenuation length to be written as (Johnson et al. ,1998):

$$\lambda = 0.5x10^6 km\left(\frac{1TeV}{E}\right)\left(\frac{3g/cm^3}{\rho}\right) \tag{19}$$

Since the neutrino attenuation length is very long, the neutrino fluence is very weakly attenuated while traversing a shield. Therefore, shielding is not an effective dose reduction tool for neutrinos.

The effective dose arising from an energy independent neutrino fluence spectrum is accomplished by performing the integration of Eq. 15 using Eqs. 16 and 17:

$$H = \int_0^{E_o} \frac{2}{E_o}\left(1 - \frac{E}{E_o}\right)\Phi\left(KE^2\right)dE = \frac{K}{6}E_o^2\Phi \tag{20}$$

where H is the annual neutrino effective dose in μSv and Φ is the total number of neutrinos per unit area that is assumed to be independent of energy (Johnson et al. ,1998).

The neutrino fluence Φ is the total number of neutrinos traversing a surface behind the shielding. The surface is governed by the divergence of the neutrino beam and the distance r from the neutrino source. The neutrino's half-divergence angle (θ) is:

$$\theta = \frac{mc^2}{E} = \frac{1}{\gamma} \approx \frac{1}{10E_o} \tag{21}$$

where $mc^2$ is the muon rest mass in MeV, E is the muon energy, θ is the opening half-angle or characteristic angle of the decay cone expressed in radians, and $E_o$ is the energy of the primary muon beam in GeV.

The neutrino fluence Φ at a given distance r from the muon decay point is just the number of neutrinos N per unit area:

$$\Phi = \frac{N}{\pi(\theta r)^2} \tag{22}$$

Combining Eqs. 20 - 22 and using the numerical value for K yields a compact form for the annual neutrino effective dose from a circular muon collider:

$$H = \frac{10^{-15}E_o^{\;2}}{6}\frac{N}{\pi(\theta r)^2} = \frac{10^{-15}E_o^{\;2}}{6}\frac{N}{\pi}\frac{(10E_o)^2}{r^2} = \frac{10^{-13}E_o^{\;4}N}{6\pi r^2}\frac{\mu Sv - cm^2}{GeV^4} \tag{23}$$

The circular muon collider neutrino effective dose of Eq. 23 has a very strong dependence on the neutrino energy.

Eq. 23 provides the neutrino effective dose assuming all muons decay at the same point. Recognizing that the muons can decay at all storage ring locations with equal probability provides a more physical description of the effective dose. For facilities such as the European Laboratory for Particle Physics (CERN), the neutrino effective dose may to be calculated as an integral over the length of the return arm (l) (Silari & Vincke, 2002) of the storage ring pointing toward the surface from d to d + l, where d is the thickness of material traversed by the neutrino beam between the end of the return arm and the surface of the earth along the direction of the return arm. The quantity d may also be described as the approximate minimum thickness of earth needed to absorb the circulating muons if beam misdirection or total beam loss occurs (i.e., the beam exits the facility). Recognizing that the muons may decay at any location along the return arm, leads to the neutrino effective dose:

$$H = \frac{10^{-13} E_o^{\ 4}}{6\pi} \int_{d}^{d+l} \frac{N}{l} \frac{dr}{r^2} = \frac{10^{-13} E_o^{\ 4} N}{6\pi l} \left( \frac{1}{d} - \frac{1}{d+l} \right) \frac{\mu Sv - cm^2}{GeV^4} \tag{24}$$

Silari & Vincke (2002) provides parameters for the planned muon facility at CERN. For a 50 GeV muon energy in the storage ring, N = $10^{21}$ muons per year decaying in the ring, a return arm length pointing toward the surface (l = $6.0 \times 10^4$ cm), and a 100 m thickness of material (d) traversed by the neutrino beam between the end of the return arm and the surface, a surface neutrino effective dose of 47 mSv/yr is predicted. Since the planned CERN design has 3 return arms, the effective dose rate at the end of one of the arms would be about 16 mSv/y (47 mSv/3). Increasing muon energy will lead to higher muon effective dose rates, additional muon shielding requirements, and will force the collider deeper underground (See Table 4, derived from Silari & Vincke, (2002).

| Muon Energy (TeV) | d (m) | L (km) | φ (mrad) | θ (μrad) |
|---|---|---|---|---|
| 1 | 100 | 36 | 5.6 | 106 |
| 2 | 100 | 36 | 5.6 | 53 |
| 5 | 200 | 51 | 8 | 21 |
| 10 | 500 | 80.5 | 12.5 | 11 |

Table 4. Geometrical Parameters for Representative Cases of Circular Muon Colliders

These results suggest that the circular muon collider be installed underground to shield the muon beam in the event the beam becomes misdirected. This required shielding is determined by the muon energy loss (Silari & Vincke, 2002):

$$\frac{dE}{dx} = 0.6 \frac{TeV}{km} \left( \frac{\rho}{3g/cm^3} \right) \tag{25}$$

When compared to muons, neutrinos have a much smaller interaction cross section. The earth shielding that completely attenuates the muons will have a negligible effect on the neutrinos. Accordingly, the neutrinos will produce a nontrivial annual effective dose at the earth's surface where the beam emerges. In order to evaluate the magnitude of this neutrino effective dose, assume the earth is a sphere, and a horizontal, circular muon collider is situated a depth d below the earth's surface. The neutrino beam exit point from the earth will be at a horizontal distance L given by Silari & Vincke (2002):

$$L = \sqrt{2dR - d^2} \approx \sqrt{2dR} \approx 36\, km \sqrt{\frac{d}{100\, m}} \tag{26}$$

where R = 6400 km is the earth's radius. Table 4 provides representative values of d and L.

In addition to d and L, a number of other relevant parameters associated with the circular collider of Eq. 26 are summarized in Table 4. In Table 4, φ is the half-angle subtended by the horizontal accelerator beam with respect to the earth's center before it exits the earth:

$$\text{Sin}\ \varphi = L / R \tag{27}$$

The functional form of Eq. 24 suggests that the calculation of neutrino effective dose from a circular muon collider is dependent of the assumed physical configuration and beam characteristics. An estimate of the neutrino effective dose for a circular muon collider can be made using Eq. 23. For comparison with Eq. 14, Eq. 23 is rewritten in terms of TeV and mSv units:

$$H = \frac{10^{-4} E_o^4 N}{6\pi r^2} \frac{mSv - cm^2}{TeV^4} \tag{28}$$

where N is the number of muon decays per year, $E_o$ is the muon energy in TeV, r is the distance from the point of muon decay in cm, and H is the annual neutrino effective dose in mSv. For consistency with the linear muon collider assumptions, $6.4 \times 10^{18}$ muon decays per year are assumed in subsequent calculations. Given the TeV muon energies and the earth shielding present, charged particle equilibrium is assumed to exist. Moreover, the neutrino beam is limited to muon neutrinos only.

The muon neutrino effective dose to fluence conversion factor is assumed to be valid at energies beyond those utilized in Cossairt et al. (1997). Given the TeV muon energies, Process D of Cossairt et al. (1997) will dominate the neutrino effective dose.

In deriving the circular muon collider effective dose relationship, a number of assumptions were made. First, the neutrino effective dose calculation assumes a 100% occupancy factor, and is an annual average based on the number of muon decays in a year. Second, the muon beam is well-collimated. In addition, the irradiated individual is (1) assumed to be within the footprint of the neutrino beam and the hadronic particle shower that results from the neutrino interactions, (2) irradiated by only one of the muon beam's decay neutrinos whose energy is one-half the total circular muon collider energy, and (3) uniformly irradiated by the neutrino and hadronic radiation types.

Table 5 summarizes the results of neutrino effective dose values as a function of distance from the muon decay location (r) for a circular muon collider. Since the facility energy is the sum of the muon and antimuon energies, a 100 TeV accelerator consists of a 50 TeV muon beam and a 50 TeV antimuon beam.

The long, thin conical radiation plumes present a radiation challenge well beyond the facility boundary. For example, a 25 TeV circular muon collider produces a neutrino effective dose of 37 mSv/y at a distance of 1500 km from the facility. Although the neutrino effective dose plume will only have a radius of 12 m at 1500 km, it presents a radiation challenge for muon collider health physicists and management. The effective dose values summarized in Table 5 have the potential to impart lethal doses to small areas. The large effective dose values and their control must be addressed in facility design and licensing.

The importance of properly characterizing offsite public effective doses is illustrated by the Fukushima Daiichi Nuclear Power Station (FDNPS) accident in Japan (Butler; 2011a, 2011b). These doses focused attention on inadequacies in the FDNPS design and licensing bases. Offsite effective doses and their profile must be carefully and credibly addressed in muon collider design and licensing evaluations.

| Accelerator Energy (TeV)[a] | H (mSv/y) at the Specified Distance (r) from the Accelerator | | | | |
|---|---|---|---|---|---|
| | 5 km | 25 km | 100 km | 1500 km | 2500 km |
| 0.1 | $8.5x10^{-4}$ | $3.4x10^{-5}$ | $2.1x10^{-6}$ | $9.4x10^{-9}$ | $3.4x10^{-9}$ |
| 2 | 140 | 5.4 | 0.34 | $1.5x10^{-3}$ | $5.4x10^{-4}$ |
| 25 | $3.3x10^{6}$ | $1.3x10^{5}$ | $8.3x10^{3}$ | 37 | 13 |
| 100 | $8.5x10^{8}$ | $3.4x10^{7}$ | $2.1x10^{6}$ | $9.4x10^{3}$ | $3.4x10^{3}$ |
| 500 | $5.3x10^{11}$ | $2.1x10^{10}$ | $1.3x10^{9}$ | $5.9x10^{6}$ | $2.1x10^{6}$ |
| 1000 | $8.5x10^{12}$ | $3.4x10^{11}$ | $2.1x10^{10}$ | $9.4x10^{7}$ | $3.4x10^{7}$ |

[a] The muon beam energy is half the accelerator energy.

Table 5. Annual Neutrino Effective Doses for a Circular Muon Collider.

Physics and cost parameters associated with 0.1, 3, 10, and 100 TeV circular muon colliders (King 1999a) are summarized in Table 6. Given current levels of technology, the collider cost will present a funding challenge as TeV muon energies are reached. In addition to funding issues, the control of radiation from the muon beams and neutrino plumes must be addressed. The feasibility of higher energy colliders will necessarily depend on technological development as well as financial support of scientific agencies.

| Accelerator Energy (TeV) | 0.1 | 3 | 10 | 100 |
|---|---|---|---|---|
| Circumference (km) | 0.35 | 6 | 15 | 100 |
| Average Magnetic Field (T) | 3.0 | 5.2 | 7.0 | 10.5 |
| Cost | Feasible | Challenging | Challenging | Problematic |

Table 6. Circular Muon Collider Physics and Cost Parameters.

As the collider energy increases, muon shielding requirements dictate a subsurface facility. The impact of locating the muon collider deeper underground with increasing accelerator energy can also be investigated. Using Eq. 28 and the data summarized in Table 4, permit the calculation of the neutrino effective dose upon its exit from the earth's surface. If the same beam properties are assumed as for the linear muon collider (i.e., $N = 6.4x10^{18}$ muon decays per year) and r = L (Table 4), then the magnitude and size of the resultant radiation plumes derived from Eq. 28 are summarized in Table 7.

| Muon Energy (TeV) [a] | d (m)[b] | L (Horizontal Distance at the Earth's Surface) (km) [c] | Beam Radius at the Earth's Surface (m) [d] | H at the Earth's Surface (mSv/y) |
|---|---|---|---|---|
| 1 | 100 | 36 | 3.6 | 2.6 |
| 2 | 100 | 36 | 1.8 | 42 |
| 5 | 200 | 51 | 1.0 | 820 |
| 10 | 500 | 80.5 | 0.8 | $5.2x10^{3}$ |
| 50 | 500 | 80.5 | 0.16 | $3.3x10^{6}$ |
| 100 | 500 | 80.5 | 0.081 | $5.2x10^{7}$ |
| 500 | 500 | 80.5 | 0.016 | $3.3x10^{10}$ |
| 1000 | 500 | 80.5 | 0.0081 | $5.2x10^{11}$ |

[a] The accelerator energy is twice the muon energy.
[b] Accelerator depth below the surface of the earth.
[c] Horizontal exit point distance from the surface of the earth.
[d] The half-divergence angle is determined from Eq. 5.

Table 7. Neutrino Effective Dose Characteristics for a Circular Muon Collider.

Although the effective dose results at the earth's surface are significant, they occur over a relatively small area. The results also assume a 100% occupancy factor for this small area, which is not likely. The magnitude of the neutrino effective dose merits significant attention and emphasis on radiation monitoring and control. For example, a 500 TeV muon beam would deliver an acute absorbed dose rate of about 1 Gy/s to a 3.2 cm diameter circle. This absorbed dose rate is sufficient to deliver a biological detriment to the body within seconds (Bevelacqua, 2010a).

Dose management controls will be similar to those enacted for direct beam exposures at conventional accelerators. Interlocks associated with beam misalignment are effective in limiting the probability that the beam is directed toward an unanticipated direction. However, additional methods to control the offsite neutrino dose must be developed because lethal exposures can occur in a very short time even though the areas involved are small. Subjecting the public to potentially lethal effective doses represents unique facility licensing challenges that must be addressed in facility safety analyses. Public perception and stakeholder involvement will be key elements in licensing TeV-PeV scale muon colliders. The need for public involvement in licensing and regulatory discussions becomes particularly important when high effective doses could result from facility operations.

## 8. Offsite effective dose considerations for muon colliders

TeV energy neutrinos do not behave according to conventional operational health physics experience at power reactors and contemporary accelerator facilities. As noted previously, neutrinos are electrically uncharged and only interact through the weak interaction. Their small, but non-zero, interaction cross section creates a unique situation in terms of the behavior of the neutrino effective dose, particularly in terms of the shape and energy dependence of their radiation profile. These properties will lead to a modification of conventional health physics dose reduction concepts when applied to planned muon colliders.

Basic radiation protection principles suggest that the effective dose at a given location is reduced if the exposure time is minimized, the distance from the source is increased, or shielding is added between the source and the point of interest (Bevelacqua, 2009, 2010a). These principles must be modified at a TeV energy muon collider. The time principle is still valid for muons and neutrinos. The neutrino and muon effective doses are reduced by decreasing the exposure time.

The distance principle is ineffective when neutrinos are involved. Since neutrinos interact very weakly, relatively long distances are not effective in significantly reducing the neutrino effective dose. In fact, the neutrino beam remains a hazard for hundreds of kilometers. However, distance will still be effective for reducing the muon effective dose.

Unlike other radiation types, shielding neutrinos increases the effective dose. The magnitude of the particle showers produced by neutrino interactions is governed by the quantity of shielding material between the neutrino beam and the point of interest. However, shielding muons is an effective dose reduction measure.

From the standpoint of TeV energy neutrino radiation, a linear muon collider has a number of advantages over circular muon colliders. Firstly, the radiation is confined to two narrow beams that can be oriented to minimize the interaction of the neutrinos. A simple dose reduction technique orients the linear accelerators at an angle such that the neutrino beams exit the accelerator above the ground. This configuration minimizes the residual neutrino interactions with the earth and man-made structures. Secondly, the spent muons can be removed from the beam following collisions or interactions before they decay into high-energy neutrinos.

## 9. Other radiation protection issues

A number of radiation protection issues associated with TeV energy muon colliders will challenge accelerator health physicists. The issues related to large neutrino effective dose values and effective neutrino dosimetry were previously noted. Before construction of a muon collider, thorough studies will be performed to define the accelerator's radiation footprint. These studies will: (1) define muon collider shielding requirements; (2) assess induced activity within the facility and the environment (e.g., air, water, and soil), including the extent of groundwater activation; (3) assess radiation streaming through facility penetrations (e.g., ventilation ducts and access points); (4) assess various accident scenarios such as loss of power or beam misdirection; and (5) assess the various pathways for liquid and airborne releases of radioactive material. Facility waste generation and decommissioning are other areas that will require evaluation.

In addition to the aforementioned radiation protection issues, the TeV energy neutrino beam will create new issues. Radiation protection concerns unique to muon colliders have been reported by Autin et al. (1999), Bevelacqua (2004), Johnson et al. (1998), Mokhov & Cossairt (1998), and Mokhov et al. (2000). These authors suggest that above about 1.5 TeV, the neutrino induced secondary radiation will pose a significant hazard even at distances on the order of tens to hundreds of kilometers. The neutrino radiation hazard presents both a physical as well as political challenge (King, 1999a).

These issues also complicate the process for locating a suitable site for a TeV energy muon collider. There are a number of potential solutions to reduce the neutrino effective dose associated with a muon collider. These include using radiation boundaries or fenced-off areas to denote areas with elevated effective dose values. Building the collider on elevated ground or at an isolated area would also minimize human exposure. Effective dose reduction measures are also available for specific muon collider configurations.

In a linear muon collider operating at the higher TeV energies, dose reduction is achieved by locating the interaction region above the earth's surface. In a circular muon collider, dose reduction is achieved by minimizing the straight sections in the ring, burying the collider deep underground to increase the distance before the neutrino beam exits the ground, and orienting the collider ring to take advantage of natural topographical features.

Orders of magnitude reductions in the neutrino effective dose are required for the muon colliders noted in this chapter (See Tables 3, 5, and 7) to meet current regulations for public exposures (ICRP, 2007). Some of the possible effective dose reduction solutions may be difficult to implement for the TeV energy muon colliders. The most feasible options for locating and operating the highest TeV energy muon collider are to either use (1) an isolated location where no one is exposed to the neutrino radiation before it exits into the atmosphere as a result of the earth's curvature, or (2) a linear muon collider constructed such that the individual muon beams collide in air well above the earth's surface.

For Option 1, the accelerator could either be constructed at an elevated location or at an isolated area. The area will need to be large, perhaps having a site boundary with a diameter greater than 100 km (King, 1999a). This size requirement restricts the available locations, and would normally require that the facility have access to the resources of an existing accelerator facility such as CERN or Fermilab. Alternatively, the facility could be located in an isolated area and scientific personnel relocated to that area with the establishment of a self-sufficient site. The final decision regarding facility location will involve funding and political considerations that are part of new facility development, licensing, and construction.

Option 2 would be technically feasible, and could be located at a smaller site. However, design considerations for both Options 1 and 2 would need to address a number of potential radiation issues associated with accelerator operation (Bevelacqua, 2008, 2009, and 2010a) that could lead to significant, unanticipated radiation levels in controlled as well as uncontrolled areas. Radiation protection issues include beam alignment errors, design errors, unauthorized changes, activation sources, and control of miscellaneous radiation sources (Bevelacqua, 2008, 2009, 2010a). These operational issues require close control because they have the potential to produce large and unanticipated effective dose values.

Beam alignment errors could direct the beam in unanticipated directions. Given the long range of the muon effective dose profile, these errors could have a significant impact on licensing and accident analysis. Beam alignment errors are caused by a variety of factors including power failures, maintenance errors, and magnet failures. Both human errors and mechanical failures lead to beam alignment issues.

Changes in the beam energy or beam current, that exceed the authorized operating envelope, lead to elevated fluence rates, the creation of unanticipated particles, or the creation of particles with higher energy than anticipated. Changes to beam parameters must be carefully evaluated for their impact on the radiation environment of the facility.

The control of secondary radiation sources, radio-frequency equipment, high-voltage power supplies, and other experimental equipment merits special attention. These sources of radiation are more difficult to control than the primary or scattered accelerator radiation because health physicists may not be aware of their existence, the experimenters may not be aware of the hazard, or the radiation source is at least partially masked by the accelerator's radiation output. These miscellaneous radiation sources will include x-rays as well as other types of radiation.

## 10. Overview of the neutrino effective dose at a tau collider

A third generation tau collider has not been evaluated. In order to provide an estimate of the effective dose consequences of a tau collider, a modification of the muon collider methodology is utilized. The decay characteristics of a tau are considerably more complex than muon decay. The muon essentially decays with a branching ratio of 100 % into a lepton and neutrinos via Eq. 1. For example, tau decays involve 119 decay modes with specified branching fractions with six modes accounting for 90% of the decays (Particle Data Group 2010). The dominant tau decay mode is:

$$\tau^- \rightarrow \pi^- + \pi^0 + \nu_\tau \ (25.51\%) \tag{29}$$

However, the negative pion dominantly decays into a muon and antimuon neutrino, and the neutral pion decays primarily into photons.

$$\tau^- \rightarrow \left(\mu^- + \bar{\nu}_\mu\right) + \left(\gamma + \gamma\right) + \nu_\tau \tag{30}$$

Subsequently, the muon decays following Eq. 1. Eq. 30 then yields:

$$\tau^- \rightarrow \left(e^- + \nu_\mu + \bar{\nu}_e + \bar{\nu}_\mu\right) + \left(\gamma + \gamma\right) + \nu_\tau \tag{31}$$

The net result of the decay is that multiple neutrinos are produced from the tau and subsequent decay of particles. The factor $\xi$ described in subsequent discussion incorporates the effects of the multiple tau decay modes and their effects on the neutrino effective dose.

Subsequent discussion assumes no annihilation of particles and antiparticles in the beam produced by the tau decay products. In addition, the narrow beam approximation is assumed.

The neutrino dose from tau decays is determined by comparing the number of neutrinos emitted from an equal number of tau and muon decays. $\xi$ defines the ratio of the number of neutrinos contributing to the tau collider to muon collider effective doses:

$$\xi = \frac{\sum_{i=1}^{N} Y_i \sum_{j=1}^{3} \left(a n_i\left(\upsilon_j\right) + b n_i\left(\bar{\upsilon}_j\right)\right)}{a n\left(\upsilon_\mu\right) + b n\left(\bar{\upsilon}_e\right)} \tag{32}$$

In the numerator of Eq. 32, i labels the various decay modes of the tau, N is the number of tau decay modes, $Y_i$ is the branching fraction of the ith tau decay mode, $n_i(\upsilon_j)$ is the number of generation j neutrinos emitted from decay mode i, and $n_i\left(\bar{\upsilon}_j\right)$ is the number of generation j antineutrinos emitted from decay mode i. In the denominator of Eq. (32), $n(\upsilon_\mu)$ is the number of muon neutrinos emitted in a muon decay, and $n\left(\bar{\upsilon}_e\right)$ is the number of antielectron neutrinos emitted in a muon decay. The j sum counts the three neutrino generations, and a and b are the cross-section factors of King (1999a) for neutrinos and antineutrinos which are 1.0 and 0.5, respectively.

The ratio of tau neutrino to muon neutrino effective doses is obtained by utilizing the value of $\xi$ and the calculated ratio of tau and muon neutrino cross-sections ($\beta$) (Jeong & Reno, 2010). The discussion is applicable to circular and linear muon and tau colliders. For equivalent accelerator operating conditions (e.g., beam energy and number of beam particle decays) and receptor conditions (e.g., distance and ambient conditions), the ratio of neutrino effective doses from a tau collider and muon collider is given by:

$$\frac{H_{\tau^-}(E)}{H_{\mu^-}(E)} = \xi \beta(E) \tag{33}$$

The results of calculations utilizing Eq. 33 are summarized in Table 8.

| Beam Energy (TeV) | Effective Dose Ratio |
|---|---|
| 0.01 | 0.39 |
| 0.1 | 1.75 |
| 1.0 | 2.16 |
| 10. | 2.23 |

Table 8. Ratio of Tau and Muon Collider Neutrino Effective Doses.

The tau collider neutrino effective doses are generally larger than those encountered in a muon collider, and the tau dose profile is also larger. The larger tau profile is demonstrated by considering Eqs. 3 and 4 for equivalent tau and muon collider configurations:

$$\frac{r\left(\tau^{-}\right)}{r\left(\mu^{-}\right)}=\frac{m_{\tau^{-}}}{m_{\mu^{-}}}=\frac{1777\,MeV}{105.7\,MeV}=16.8 \tag{34}$$

Using Eq. 34 and the Table 7 results for circular tau collider conditions, the neutrino effective dose profile radius at the earth's surface is 60.5, 30.2, 16.8, and 13.4 m for 1, 2, 5, and 10 TeV beams. These affected areas and associated effective doses suggest that the tau collider is a more significant radiation hazard than the muon collider. Therefore, larger effective doses and affected areas are anticipated during tau collider operations.

An improved calculation of the neutrino effective dose from a tau collider requires a better specification of neutrino properties. For example, previous calculations were based on the Standard Model assumption that neutrinos have zero mass. Neutrino masses can be calculated assuming the alternative gauge group $SU(2)_L \otimes SU(2)_R \otimes U(1)$ instead of the Standard Model $SU(2)_L \otimes U(1)$. This gauge group leads to a neutrino generation i mass:

$$m_i = \frac{M_i^2}{g\, m_{W_R}} \tag{35}$$

where $M_i$ is the generation i lepton mass (e, μ, and τ), $W_R$ is the right-handed W boson mass (≥ 300 GeV), and g is a coupling constant with a value of 0.585 (Mohapatra & Senjanović, 1980). Using these values in Eq. 35 leads to electron, muon, and tau neutrino upper bound masses of 1.5 eV, 64 keV, and 18 MeV, respectively. These masses affect the input values used to calculate the neutrino effective dose in Eqs. 14 and 23. As an alternative, better cross-section data and dose conversion factors would refine the neutrino effective dose.

## 11. Conclusions

Neutrino radiation will be a health physics issue and design constraint for muon colliders, particularly at TeV energies. TeV energy muon colliders will require careful site selection and the neutrino effective dose may dictate that these machines be constructed in isolated areas. With the operation of TeV energy muon colliders, the neutrino effective dose can no longer be neglected. Neutrino detection, neutrino dosimetry, and the determination of the neutrino effective dose will no longer be academic exercises, but will become operational health physics concerns. Keeping public and occupational neutrino effective doses below regulatory limits will require careful and consistent application of dose reduction methods.

When compared to muon colliders, initial scooping calculations for tau colliders suggest that higher effective doses and affected areas will result from their operation. Although, the tau collider calculations are initial estimates, they suggest that significant radiation challenges are also presented by these machines.

## 12. References


Autin, B; Blondel, A. & Ellis, J. (1999). Prospective Study of Muon Storage Rings at CERN, CERN 99-02, European Laboratory for Particle Physics, Geneva, Switzerland

Bevelacqua, J. (2004). Muon Colliders and Neutrino Dose Equivalents: ALARA Challenges for the 21st Century, *Radiation Protection Management*, Vol.21, No. 4, pp. 8-30.

Bevelacqua, J. (2008). *Health Physics in the 21st Century*, Wiley-VCH, ISBN 9783527408221, Weinheim, Germany

Bevelacqua, J. (2009). *Contemporary Health Physics: Problems and Solutions* (Second Edition), ISBN 9783527408245, Weinheim, Germany

Bevelacqua, J. (2010a). *Basic Health Physics: Problems and Solutions* (Second Edition), ISBN 9783527408238, Weinheim, Germany

Bevelacqua, J. (2010b). Standard Model of Particle Physics-A Health Physics Perspective, *Health Physics*, Vol.99, No.5, pp. 613-623

Butler, D. (2011a). Radioactivity Spreads in Japan, *Nature*, Vol.471, No.7340, pp. 555-556

Butler, D. (2011b). Fukushima Health Risks Scrutinized, *Nature*, Vol.472, No.7341, pp. 13-14

Cottingham, W. & Greenwood, D. (2007). *An Introduction to the Standard Model of Particle Physics* (Second Edition), Cambridge University Press, ISBN 9780521852494, Cambridge, UK

Cossairt, J.; Grossman, N. & Marshall, E. (1996). Neutrino Radiation Hazards: A Paper Tiger, *Fermilab-Conf-96/324*, Accessed on July 11, 2011, Available from: <http://lss.fnal.gov/archive/1996/conf/Conf-96-324.pdf>

Cossairt, J.; Grossman, N. & Marshall, E. (1997). Assessment of Dose Equivalent due to Neutrinos, *Health Physics*, Vol.73, No.6, 894-898.

Cossairt, J. & Marshall, E. (1997). Comment on "Biological Effects of Stellar Collapse Neutrinos, *Physical Review Letters*, Vol.78, No.7, pp.1394.

Collar, J. (1996). Biological Effects of Stellar Collapse Neutrinos, *Physical Review Letters*, Vol.76, No.6, pp. 999-1002

Geer, S. (2010). From Neutrino Factory to Muon Collider, *FERMILAB-CONF-10-024-APC*, Accessed on July 14, 2011, Available from: < http://arxiv.org/abs/1006.0923>

Griffiths, D. (2008). *Introduction to Elementary Particle Physics* (Second Edition), Wiley-VCH, ISBN 9783527406012, Weinheim, Germany

ICRP Report No. 60. (1991). *1990 Recommendations of the International Commission on* Radiological Protection, Elsevier, Amsterdam

ICRP Report No. 107. (2007). *The 2007 Recommendations of the International Commission on Radiological Protection*, Elsevier, Amsterdam

Jeong, Y. & Reno, M. (2010). Tau neutrino and antineutrino cross sections, Accessed on July 12, 2011, Available from: <http://arxiv.org/PS_cache/arxiv/pdf/1007/1007.1966v1.pdf>

Johnson, C.; Rolandi, G. & Silari, M. (1998). Radiological Hazard due to Neutrinos from a Muon Collider, *Internal Report CERN/TIS-RP/IR/98*, European Laboratory for

Particle Physics (CERN), Geneva, Switzerland (1998). Accessed on July 12, 2011, Available from: <http://www.physics.princeton.edu/mumu/johnson/neutrino.pdf>

King, B. (1999a). Studies for Muon Collider Parameters at Center-of-Mass Energies of 10 TeV and 100 TeV, Brookhaven National Laboratory, Accessed on July 25, 2011, Available from: <http://arxiv.org/PS_cache/physics/pdf/9908/9908018v1.pdf >

King, B. (1999b). Neutrino Radiation Challenges and Proposed Solutions for Many-TeV Muon Colliders, *Proc. HEMC'99 Workshop – Studies on Colliders and Collider Physics at the Highest Energies: Muon Colliders at 10 TeV to 100 TeV*, Montauk, NY, September 1999 Accessed on July 25, 2011, Available from: <http://nfmcc-docdb.fnal.gov/cgi-bin/RetrieveFile?docid=119&version=1&filename=muc0119.ps.gz>

King, B. (1999c). Neutrino Physics at Muon Colliders, Brookhaven National Laboratory, Accessed on July 19, 2011, Available from: <http://arxiv.org/PS_cache/hep-ex/pdf/9907/9907035v1.pdf>

Kuno, Y. (2009). Project X Workshop Summary, *Muon Collider Physics Workshop*, Fermi National Laboratory, Batavia, IL, 10 -12 November, 2009, Available from: <https://indico.fnal.gov/getFile.py/access?contribId=78&sessionId=0&resId=0&materialId=slides&confId=2855>

Mohapatra, R. & Senjanović, G. (1980). Neutrino Mass and Spontaneous Parity Nonconservation, *Physical Review Letters*, Vol.44, No.14, pp. 912-915

Mokhov, N. & Cossairt, J. (1998). Radiation Studies at Fermilab, *Proceedings of the Fourth Workshop on Simulating Accelerator Radiation Environments (SARE4)*, Knoxville, TN, 14 - 16 September, 1998, Accessed on July 25, 2011, Available from: <http://lss.fnal.gov/archive/1998/conf/Conf-98-384.pdf >

Mokhov, N.; Striganov, S. & van Ginneken, A. (2000). Muons and Neutrinos at High Energy Accelerators, *FERMILAB-Conf-00/182*, Accessed on July 11, 2011, Available from: <http://lss.fnal.gov/archive/2000/conf/Conf-00-182.pdf>

NCRP Report No. 144. (2003). *Radiation Protection for Particle Accelerator Facilities*, National Council on Radiation Protection and Measurements, ISBN 0929600770, Bethesda, MD

Particle Data Group. (2010). Review of Particle Properties, *Journal of Physics G*, Vol.37, No.7A, pp. 1-1422.

Quigg, C. (1997). Neutrino Interaction Cross Sections, *FERMILAB-Conf-97/158-T*, Accessed on July 11, 2011, Available from: < http://lss.fnal.gov/archive/1997/conf/Conf-97-158-T.pdf>

Silari, M. & Vincke, H. (2002). Neutrino Radiation Hazard at the Planned CERN Neutrino Factory, *Technical Note TIS-RP/TN/2002-01*, Accessed on July 12, 2011, Available from: <http://slap.web.cern.ch/slap/NuFact/NuFact/nf105.pdf>

Zimmerman, F. (1999). Final Focus Challenges for Muon Colliders at Highest Energies, *Proc. HEMC'99 Workshop – Studies on Colliders and Collider Physics at the Highest Energies: Muon Colliders at 10 TeV to 100 TeV*, Montauk, NY, September 1999, Accessed on July 25, 2011, Available from: <//cdsweb.cern.ch/record/420774/files/sl-1999-077.ps.gz>

Zisman, M. (2011). R&D Toward a Neutrino Factory and Muon Collider, LBNL-4494E, Lawrence Berkeley National Laboratory, Berkeley, CA, Accessed on July 25, 2011, Available from: <http://escholarship.org/uc/item/43p7z0v1>